Institute for Mathematics and its Applications
IMA

The **Institute for Mathematics and its Applications** was established by a grant from the National Science Foundation to the University of Minnesota in 1982. The IMA seeks to encourage the development and study of fresh mathematical concepts and questions of concern to the other sciences by bringing together mathematicians and scientists from diverse fields in an atmosphere that will stimulate discussion and collaboration.

The IMA Volumes are intended to involve the broader scientific community in this process.

Avner Friedman, Director

Robert Gulliver, Associate Director

* * * * * * * * * *

IMA ANNUAL PROGRAMS

1982–1983	Statistical and Continuum Approaches to Phase Transition
1983–1984	Mathematical Models for the Economics of Decentralized Resource Allocation
1984–1985	Continuum Physics and Partial Differential Equations
1985–1986	Stochastic Differential Equations and Their Applications
1986–1987	Scientific Computation
1987–1988	Applied Combinatorics
1988–1989	Nonlinear Waves
1989–1990	Dynamical Systems and Their Applications
1990–1991	Phase Transitions and Free Boundaries
1991–1992	Applied Linear Algebra
1992–1993	Control Theory and its Applications
1993–1994	Emerging Applications of Probability
1994–1995	Waves and Scattering
1995–1996	Mathematical Methods in Material Science
1996–1997	Mathematics of High Performance Computing
1997–1998	Emerging Applications of Dynamical Systems
1998–1999	Mathematics in Biology
1999–2000	Reactive Flows and Transport Phenomena

Continued at the back

Lorenz T. Biegler Thomas F. Coleman
Andrew R. Conn Fadil N. Santosa
Editors

Large-Scale Optimization with Applications

Part I: Optimization in Inverse Problems and Design

With 87 Illustrations

Springer

Lorenz T. Biegler
Chemical Engineering Department
Carnegie Mellon University
Pittsburgh, PA 15213, USA

Thomas F. Coleman
Computer Science Department
Cornell University
Ithaca, NY 14853-0001, USA

Andrew R. Conn
Thomas J. Watson Research Center
P.O. Box 218
Yorktown Heights, NY 10598, USA

Fadil N. Santosa
School of Mathematics
University of Minnesota
Minneapolis, MN 55455, USA

Series Editors:
Avner Friedman
Robert Gulliver
Institute for Mathematics and its
Applications
University of Minnesota
Minneapolis, MN 55455, USA

Mathematics Subject Classifications (1991): 65Kxx, 90Cxx, 93-XX, 90Bxx, 68Uxx, 92Exx, 92E10, 35R30, 86A22, 73Kxx, 78A40, 78A45

Library of Congress Cataloging-in-Publication Data
Large-scale optimization with applications / Lorenz T. Biegler . . . [et al.].
p. cm. – (The IMA volumes in mathematics and its applications ; 92–94)
Presentations from the IMA summer program held July 10–28, 1995.
Includes bibliographical references.
Contents: pt. 1. Optimization in inverse problems and design – pt. 2. Optimal design and control – pt. 3. Molecular structure and optimization.
ISBN 0-387-98286-8 (pt. 1 : alk. paper). – ISBN 0-387-98287-6 (pt. 2 : alk. paper). – ISBN 0-387-98288-4 (pt. 3 : alk. paper)
1. Mathematical optimization – Congresses. 2. Programming (Mathematics) – Congresses. 3. Inverse problems (Differential equations) – Congresses. 4. Engineering design – Congresses. 5. Molecular structure – Mathematical models – Congresses. I. Biegler, Lorenz T. II. Series: IMA volumes in mathematics and its applications ; v. 92–94.
QA402.5.L356 1997
500.2'01'5193 – dc21 97-22879

Printed on acid-free paper.

Production managed by Karina Mikhli; manufacturing supervised by Jacqui Ashri.
Camera-ready copy prepared by the IMA.
Printed and bound by Braun-Brumfield, Inc., Ann Arbor, MI.
Printed in the United States of America.

9 8 7 6 5 4 3 2 1

ISBN 0-387-98286-8 Springer-Verlag New York Berlin Heidelberg SPIN 10632948

FOREWORD

This IMA Volume in Mathematics and its Applications

LARGE-SCALE OPTIMIZATION WITH APPLICATIONS, PART I: OPTIMIZATION IN INVERSE PROBLEMS AND DESIGN

is one of the three volumes based on the proceedings of the 1995 IMA three-week Summer Program on "Large-Scale Optimization with Applications to Inverse Problems, Optimal Control and Design, and Molecular and Structural Optimization." The other two related proceedings appeared as Volume 93: Large-Scale Optimization with Applications, Part II: Optimal Design and Control, and Volume 94: Large-Scale Optimization with Applications, Part III: Molecular Structure and Optimization.

We would like to thank Lorenz T. Biegler, Thomas F. Coleman, Andrew R. Conn, and Fadil N. Santosa for their excellent work as organizers of the meetings and for editing the proceedings.

We also take this opportunity to thank the National Science Foundation (NSF), the Department of Energy (DOE), and the Alfred P. Sloan Foundation, whose financial support made the workshops possible.

Avner Friedman

Robert Gulliver

GENERAL PREFACE
LARGE-SCALE OPTIMIZATION
WITH APPLICATIONS, PARTS I, II, AND III

There has been enormous progress in large-scale optimization in the past decade. In addition, the solutions to large nonlinear problems on moderate workstations in a reasonable amount of time are currently quite possible. In practice for many applications one is often only seeking improvement rather than assured optimality (a reason why local solutions often suffice). This fact makes problems that at first sight seem impossible quite tractable. Unfortunately and inevitably most practitioners are unaware of some of the most important recent advances. By the same token, most mathematical programmers have only a passing knowledge of the issues that regularly arise in the applications.

It can still be fairly said that the vast majority of large-scale optimization modeling that is carried out today is based on linearization, undoubtedly because linear programming is well understood and known to be effective for very large instances. However, the world is not linear and accurate modeling of physical and scientific phenomena frequently leads to large-scale nonlinear optimization.

A three-week workshop on Large-Scale Optimization was held at the IMA from July 10 to July 28, 1995 as part of its summer program. These workshops brought together some of the world's leading experts in the areas of optimization, inverse problems, optimal design, optimal control and molecular structures. The content of these volumes represent a majority of the presentations at the three workshops. The presentations, and the subsequent articles published here are intended to be useful and accessible to both the mathematical programmers and those working in the applications. Perhaps somewhat optimistically, the hope is that the workshops and the proceedings will also initiate some long-term research projects and impart to new researchers the excitement, vitality and importance of this kind of cooperation to the applications and to applied mathematics.

The format of the meetings was such that we tried to have an invited speaker with expertise in an application of large-scale optimization describe the problem characteristics in the application, current solution approaches and the difficulties that suggest areas for future research. These presentations were complemented by an optimization researcher whose object was to present recent advances related to the difficulties associated with the topic (*e.g.*, improved methods for nonlinear optimization, global optimization, exploiting structure). One difficulty was that although it is possible (but perhaps not desirable) to isolate a particular application, the optimization methods tended to be intertwined in all of the topics.

These Proceedings include the same mix of details of the application, overview of the optimization techniques available, general discussions of the difficulties and areas for future research.

We are grateful to all the help we had from the IMA, and in particular we would like to single out Avner Friedman, Robert Gulliver and Patricia Brick whose help and support was invaluable. Patricia Brick is especially acknowledged for all of her efforts typesetting and assembling these volumes. The speakers, the attendees and the diligent reviewers of the submitted papers also deserve our acknowledgment; after all, without them there would be no proceedings. Finally we would like to thank those agencies whose financial support made the meeting possible: The National Science Foundation, the Department of Energy, and the Alfred P. Sloan Foundation.

Lorenz T. Biegler

Thomas F. Coleman

Andrew R. Conn

Fadil N. Santosa

PREFACE FOR PART I

Inverse problems and optimal design have come of age as a consequence of the availability of better, more accurate, and more efficient, simulation packages. Many of these simulators, which can run on small workstations, can capture the complicated behavior of the physical systems they are modeling, and have become commonplace tools in engineering and science. There is a great desire to use them as part of a process by which measured field data are analyzed or by which design of a product is automated. A major obstacle in doing precisely this is that one is ultimately confronted with a large-scale optimization problem.

The problem of determining the parameters of a physical system from measured field data is the subject of inverse problems. In optimal design, one is interested in determining parameters of a system so that its behavior comes close to meeting certain design objectives. Both types of problems are naturally posed as optimization or mathematical programming problems.

A common feature of these problems, from the perspective of optimization, is that function evaluations (simulations) are relatively costly. Additionally, derivative information about the functions is often not readily available, and the problems are frequently ill-posed. Moreover, the cost (or objective) function may have multiple local minima or not be differentiable.

This volume contains expository articles on both inverse problems and design problems formulated as optimization. Each paper describes the physical problem in some detail and is meant to be accessible to researchers in optimization as well as those who work in applied areas where optimization is a key tool. What emerges in the presentations is that there are features about the problem that must be taken into account in posing the objective function, and in choosing an optimization strategy. In particular there are certain structures peculiar to the problems that deserve special treatment, and there is ample opportunity for parallel computation.

Lorenz T. Biegler

Thomas F. Coleman

Andrew R. Conn

Fadil N. Santosa

CONTENTS

Large-Scale Optimization with Applications, Part I: Optimization in Inverse Problems and Design

CONTENTS OF PART II: OPTIMAL DESIGN AND CONTROL

CONTENTS OF PART III: MOLECULAR STRUCTURE AND OPTIMIZATION

SPACE MAPPING OPTIMIZATION FOR ENGINEERING DESIGN

JOHN W. BANDLER*†, RADEK M. BIERNACKI*†, SHAOHUA CHEN*†, RONALD H. HEMMERS†, AND KAJ MADSEN‡

Abstract. This contribution describes the Space Mapping (SM) technique which is relevant to engineering optimization. The SM technique utilizes different models of the same physical object. Similarly to approximation, interpolation, variable-complexity, response surface modeling, surrogate models, and related techniques, the idea is to replace computationally intensive simulations of an accurate model by faster though less accurate evaluations of another model. In contrast, SM attempts to establish a mapping between the input parameter spaces of those models. This mapping allows us to redirect the optimization-related calculations to the fast model while preserving the accuracy and confidence offered by a few well-targeted evaluations of the accurate model. The SM technique has been successfully applied in the area of microwave circuit design. In principle, however, it is applicable to a wide range of problems where models of different complexity and computational intensity are available, although insight into engineering modeling in the specific application area might be essential.

Key words. Space Mapping, Design Optimization, Engineering Modeling.

1. Introduction. Advances in computing technology provide engineering designers with more and more sophisticated simulation tools that can more and more accurately predict the behaviour of more and more complex objects. For design optimization, this means that the simulation times rather than the optimization algorithms become a true bottleneck and limiting factors for various applications. For such applications the scale of optimization is measured not in terms of the number of variables, but rather in terms of real time needed to complete it. A small electronic circuit may require weeks of simulation time and only a small fraction of a second for the optimizer to determine the next point.

This paper describes the concept of Space Mapping (SM) [2-8]. The SM technique exploits a mathematical link between input parameters of different engineering models of the same physical object: (1) a computationally efficient (fast) model which may lack the desired accuracy, and (2) an accurate but CPU-intensive model. Traditionally, there exists a number of engineering models of different types and levels of complexity to choose

* Optimization Systems Associates Inc., P.O. Box 8083, Dundas, Ontario, Canada L9H 5E7. The work of the first three authors was supported in part by Optimization Systems Associates Inc.

† The Simulation Optimization Systems Research Laboratory and Department of Electrical and Computer Engineering, McMaster University, Hamilton, Ontario, Canada L8S 4L7. The work of the first four authors was supported in part by the Natural Sciences and Engineering Research Council of Canada under Grants OGP0007239, OGP0042444, STR0167080 and through the Micronet Network of Centres of Excellence.

‡ Institute of Mathematical Modeling, Technical University of Denmark, DK-2800 Lyngby, Denmark.

from. In the field of electronic circuits this may include equivalent circuit models, ideal and detailed empirical models, electromagnetic (EM) field theory based models, hybrid models [20], and even computational utilization of actual hardware measurements.

As a fast model we want to use an engineering model that is capable of reasonably emulating the behaviour of the actual physical object, for example of generating a band-selective response of a filter. Such a model may be an otherwise good empirical model, however, the parameter values might be outside their recommended validity ranges. It may also be the same simulator which is used as the accurate model, e.g., an electromagnetic field solver, but with a coarser grid size than required to produce accurate results [4,5].

Similarly to various approximation and interpolation [10,11], variable-complexity [13,15,16], response surface modeling [19], surrogate models [23], and related techniques [21,22], the idea of SM is to replace computationally intensive simulations of an accurate model by faster though less accurate evaluations of another model. In contrast to those methods, SM attempts to establish a mapping between the input parameter spaces of those models, instead of trying to refine and validate the fast model using the same parameter values as for the accurate model. The underlying assumption is that the inaccuracy of the fast model can be somehow compensated for by adjusting the model parameter values.

The SM concept facilitates the demanding requirements of, otherwise CPU-prohibitive, design optimization within a practical time frame. This is accomplished by redirecting the optimization-related calculations to the fast model while preserving the accuracy and confidence offered by a few well-targeted evaluations of the slow accurate model.

There are two phases in SM. In *Phase 1*, optimization is performed using the fast model to obtain its optimal performance. In *Phase 2*, a mapping between the input parameter spaces of the two models considered is iteratively established. Techniques such as least-squares or quasi-Newton steps are used to accomplish this. A distinct auxiliary optimization (parameter extraction), also using the fast model, must be invoked in each iteration of *Phase 2*. This parameter extraction is used to determine the parameters of the fast model such that its response(s) match those of the reference response(s) obtained from an evaluation of the accurate model. The uniqueness of the parameter extraction process is of utmost importance to the success of SM.

The SM technique has been successfully applied in the area of microwave circuit design. In principle, however, it is applicable to a wide variety of problems where models of different complexity and computational intensity are available, although insight into engineering modeling in the specific application area might be essential.

Our presentation describes the theoretical formulation of SM followed by a practical engineering example: the design of a high-temperature superconducting (HTS) microstrip filter. In this particular case we consider an equivalent empirical circuit model as the fast mode. As the accurate model, we employ an extremely CPU-intensive model based on solving electromagnetic field equations.

2. Space mapping optimization.

2.1. Theory. Let the behaviour of a system be described by models in two spaces: the optimization space, denoted by $\boldsymbol{X}_{os}$, and the EM (or validation) space, denoted by $\boldsymbol{X}_{em}$. We represent the optimizable model parameters in these spaces by the vectors $\boldsymbol{x}_{os}$ and $\boldsymbol{x}_{em}$, respectively. We assume that $\boldsymbol{X}_{os}$ and $\boldsymbol{X}_{em}$ have the same dimensionality, i.e., $\boldsymbol{x}_{os} \in \mathbb{R}^n$ and $\boldsymbol{x}_{em} \in \mathbb{R}^n$, but may not represent the same parameters. We assume that the $\boldsymbol{X}_{os}$-space model responses, denoted by $\boldsymbol{R}_{os}(\boldsymbol{x}_{os})$, are much faster to calculate but less accurate than the $\boldsymbol{X}_{em}$-space model responses, denoted by $\boldsymbol{R}_{em}(\boldsymbol{x}_{em})$.

The key idea behind SM optimization is the generation of an appropriate mapping, $\boldsymbol{P}$, from the $\boldsymbol{X}_{em}$ space to the $\boldsymbol{X}_{os}$ space,

$$\boldsymbol{x}_{os} = \boldsymbol{P}(\boldsymbol{x}_{em}) \tag{2.1}$$

such that

$$\boldsymbol{R}_{os}(\boldsymbol{P}(\boldsymbol{x}_{em})) \approx \boldsymbol{R}_{em}(\boldsymbol{x}_{em}). \tag{2.2}$$

We assume that such a mapping exists and is one-to-one within some local modeling region encompassing our SM solution. We also assume that, based on (2.2), for a given $\boldsymbol{x}_{em}$ its image $\boldsymbol{x}_{os}$ in (2.1) can be found by a suitable parameter extraction procedure, and that this process is unique.

We initially perform optimization entirely in $\boldsymbol{X}_{os}$ to obtain the optimal solution $\boldsymbol{x}^*_{os}$, for instance in the minimax sense [9], and subsequently use SM to find the mapped solution $\overline{\boldsymbol{x}}_{em}$ in $\boldsymbol{X}_{em}$ as

$$\overline{\boldsymbol{x}}_{em} = \boldsymbol{P}^{-1}(\boldsymbol{x}^*_{os}) \tag{2.3}$$

once the mapping (2.1) is established. We designate $\overline{\boldsymbol{x}}_{em}$ as the SM solution instead of $\boldsymbol{x}^*_{em}$ since the mapped solution represents only an approximation to the true optimum in $\boldsymbol{X}_{em}$. The mapping is established through an iterative process. We begin with a set of m $\boldsymbol{X}_{em}$-space model base points

$$\boldsymbol{B}_{em} = \{\boldsymbol{x}^{(1)}_{em}, \boldsymbol{x}^{(2)}_{em}, \ldots, \boldsymbol{x}^{(m)}_{em}\}. \tag{2.4}$$

These initial m base points are selected in the vicinity of a reasonable candidate for the $\boldsymbol{X}_{em}$-space model solution. For example, if $\boldsymbol{x}_{em}$ and $\boldsymbol{x}_{os}$ consist of the same physical parameters, then the set $\boldsymbol{B}_{em}$ can be chosen as

$$\boldsymbol{x}_{em}^{(1)} = \boldsymbol{x}_{os}^{*} \tag{2.5}$$

with the remaining $m-1$ base points chosen arbitrarily by perturbation as

$$\boldsymbol{x}_{em}^{(i)} = \boldsymbol{x}_{em}^{(1)} + \Delta\boldsymbol{x}_{em}^{(i-1)}, \qquad i = 2, 3, \ldots, m. \tag{2.6}$$

For instance, for a linear mapping, one can select perturbations along the axes. Once the set $\boldsymbol{B}_{em}$ is chosen, we perform EM analyses at each base point to obtain the $\boldsymbol{X}_{em}$-space model responses $\boldsymbol{R}_{em}(\boldsymbol{x}_{em}^{(i)})$ for $i = 1, 2, \ldots, m$. This is followed by parameter extraction optimization in $\boldsymbol{X}_{os}$ to obtain the corresponding set of m $\boldsymbol{X}_{os}$-space model base points

$$\boldsymbol{B}_{os} = \{\boldsymbol{x}_{os}^{(1)}, \boldsymbol{x}_{os}^{(2)}, \ldots, \boldsymbol{x}_{os}^{(m)}\}. \tag{2.7}$$

The parameter extraction process is carried out by the following m optimizations:

$$\underset{\boldsymbol{x}_{os}^{(i)}}{\text{minimize}} \; ||\boldsymbol{R}_{os}(\boldsymbol{x}_{os}^{(i)}) - \boldsymbol{R}_{em}(\boldsymbol{x}_{em}^{(i)})|| \tag{2.8}$$

for $i = 1, 2, \ldots, m$ where $|| \cdot ||$ indicates a suitable norm. The additional $m-1$ points apart from $\boldsymbol{x}_{em}^{(1)}$ are required merely to establish full-rank conditions leading to the initial approximation of the mapping (denoted by $\boldsymbol{P}_0$ — its exact construction is explained later). At the jth iteration, both sets may be expanded to contain, in general, m_j points which are used to establish the updated mapping $\boldsymbol{P}_j$. Since the analytical form of $\boldsymbol{P}$ is not available, we use the current approximation $\boldsymbol{P}_j$ to estimate $\overline{\boldsymbol{x}}_{em}$ in (2.3), i.e.,

$$\boldsymbol{x}_{em}^{(m_j+1)} = \boldsymbol{P}_j^{-1}(\boldsymbol{x}_{os}^{*}). \tag{2.9}$$

The process continues iteratively until the termination condition

$$||\boldsymbol{R}_{os}(\boldsymbol{x}_{os}^{*}) - \boldsymbol{R}_{em}(\boldsymbol{x}_{em}^{(m_j+1)})|| \le \epsilon \tag{2.10}$$

is satisfied, where ϵ is acceptably small. If so, $\boldsymbol{P}_j$ is our desired $\boldsymbol{P}$. If not, the set of base points in $\boldsymbol{B}_{em}$ is augmented by $\boldsymbol{x}_{em}^{(m_j+1)}$ and correspondingly, $\boldsymbol{x}_{os}^{(m_j+1)}$ determined by (2.8) augments the set of base points in $\boldsymbol{B}_{os}$. Upon termination, we set $\overline{\boldsymbol{x}}_{em} = \boldsymbol{x}_{em}^{(m_j+1)} = \boldsymbol{P}_j^{-1}(\boldsymbol{x}_{os}^{*})$ as the SM solution. This process is illustrated graphically in Fig. 1.

We define each of the transformations $\boldsymbol{P}_j$ as a linear combination of some predefined and fixed fundamental functions

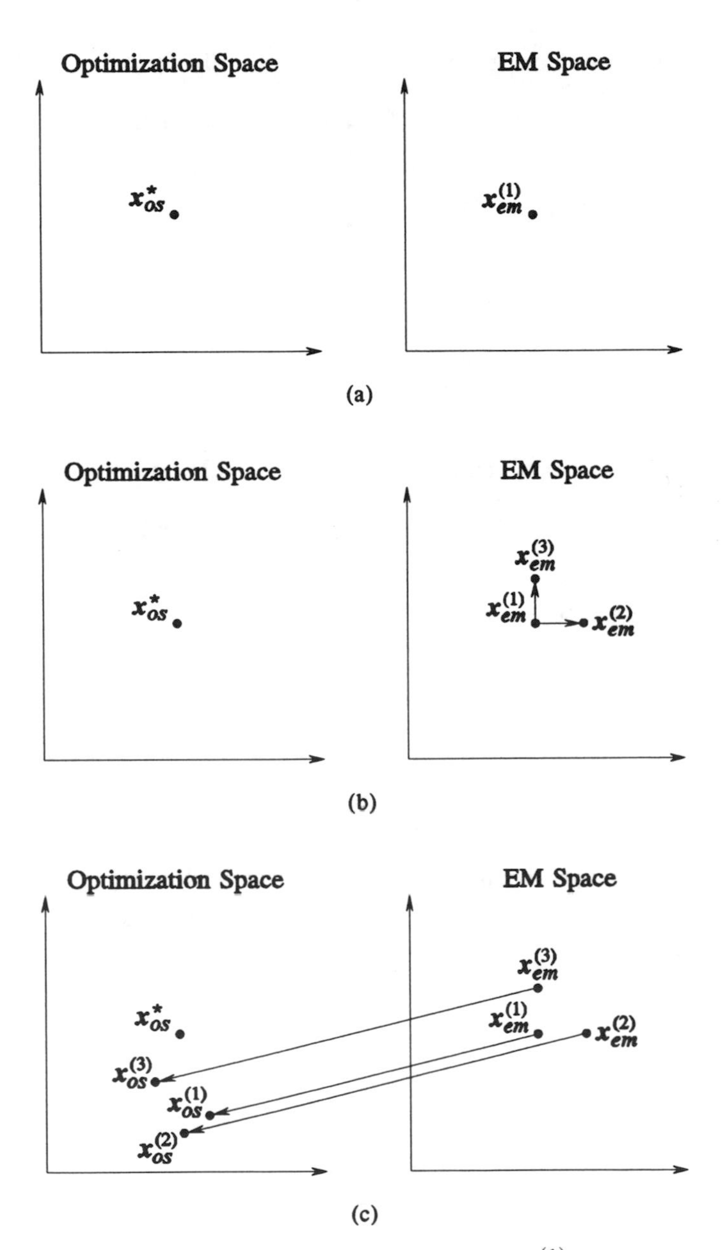

FIG. 1. *Illustration of Space Mapping optimization: (a) set $\boldsymbol{x}_{em}^{(1)} = \boldsymbol{x}_{os}^*$, assuming $\boldsymbol{x}_{em}$ and $\boldsymbol{x}_{os}$ represent the same physical parameters, (b) generate additional base points around $\boldsymbol{x}_{em}^{(1)}$, (c) perform $\boldsymbol{X}_{os}$-space model parameter extractions according to (2.8).*

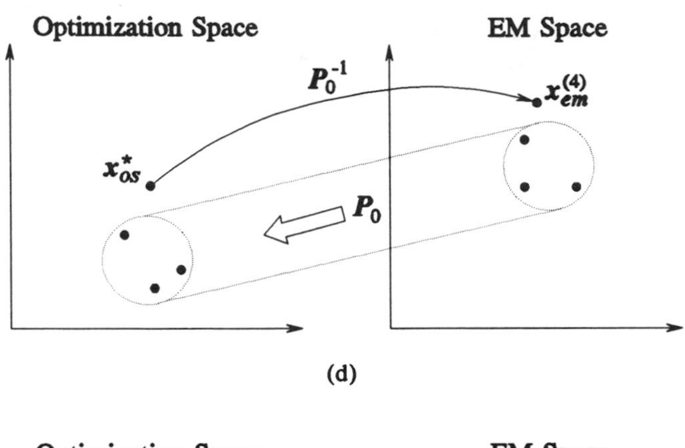

(d)

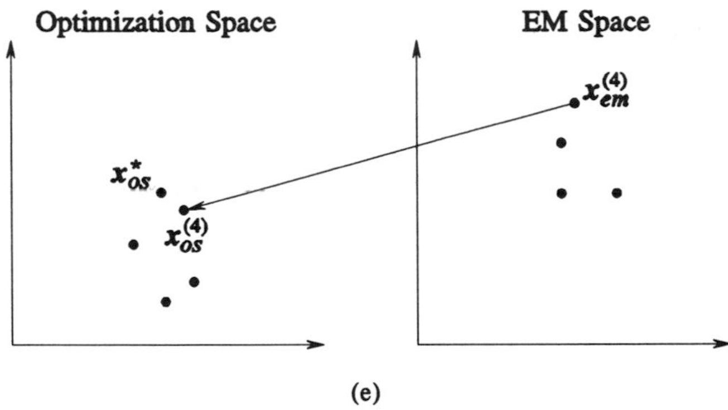

(e)

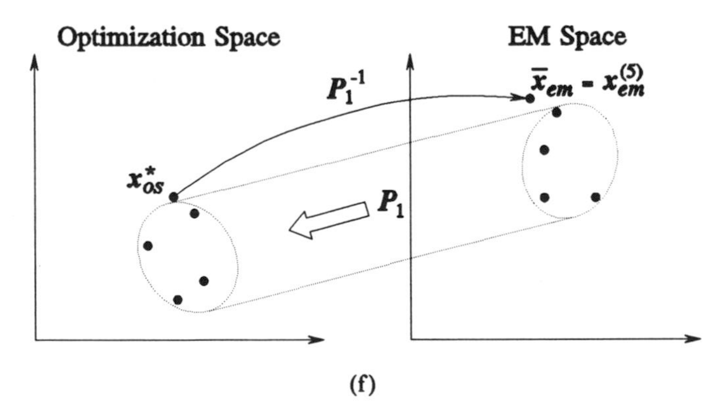

(f)

FIG. 1. *Illustration of Space Mapping optimization (cont.): (d) use the inverse mapping to obtain $\boldsymbol{x}_{em}^{(4)}$, (e) perform $\boldsymbol{X}_{os}$-space model parameter extraction to obtain $\boldsymbol{x}_{os}^{(4)}$, (f) apply the updated inverse mapping to obtain the SM solution $\overline{\boldsymbol{x}}_{em} = \boldsymbol{x}_{em}^{(5)}$, assuming $||\boldsymbol{R}_{os}(\boldsymbol{x}_{os}^*) - \boldsymbol{R}_{em}(\boldsymbol{x}_{em}^{(5)})|| \leq \epsilon$.*

$$\widehat{f}_1(\boldsymbol{x}_{em}), \widehat{f}_2(\boldsymbol{x}_{em}), \widehat{f}_3(\boldsymbol{x}_{em}), \ldots, \widehat{f}_t(\boldsymbol{x}_{em}) \tag{2.11}$$

such that

$$x_{os_i} = \sum_{s=1}^{t} a_{is} \widehat{f}_s(\boldsymbol{x}_{em}) \tag{2.12}$$

or, in matrix form

$$\boldsymbol{x}_{os} = \boldsymbol{P}_j(\boldsymbol{x}_{em}) = \boldsymbol{A}_j \widehat{\boldsymbol{f}}(\boldsymbol{x}_{em}) \tag{2.13}$$

where $\boldsymbol{A}_j$ is a $n \times t$ matrix, $\widehat{\boldsymbol{f}}(\boldsymbol{x}_{em})$ is a t-dimensional column vector of fundamental functions.

Consider the mapping $\boldsymbol{P}_j$ for all points in the sets $\boldsymbol{B}_{em}$ and $\boldsymbol{B}_{os}$. Expanding (2.13) gives

$$\left[\boldsymbol{x}_{os}^{(1)}\, \boldsymbol{x}_{os}^{(2)} \ldots \boldsymbol{x}_{os}^{(m_j)}\right] = \boldsymbol{A}_j \left[\widehat{\boldsymbol{f}}(\boldsymbol{x}_{em}^{(1)})\, \widehat{\boldsymbol{f}}(\boldsymbol{x}_{em}^{(2)}) \ldots \widehat{\boldsymbol{f}}(\boldsymbol{x}_{em}^{(m_j)})\right] \tag{2.14}$$

where $m_j \geq t$.

The simplest choice, frequently adequate for local problems, is to consider a linear mapping. In this particular case, $\widehat{\boldsymbol{f}}(\boldsymbol{x}_{em})$ in (2.11) contains the $n+1$ linear functions: $1, x_{em_1}, x_{em_2}, \ldots, x_{em_n}$. Hence, (2.14) can be written as

$$\begin{aligned} \left[\boldsymbol{x}_{os}^{(1)}\, \boldsymbol{x}_{os}^{(2)} \ldots \boldsymbol{x}_{os}^{(m_j)}\right] &= \boldsymbol{Q}_j \left[\boldsymbol{x}_{em}^{(1)}\, \boldsymbol{x}_{em}^{(2)} \ldots \boldsymbol{x}_{em}^{(m_j)}\right] + [\boldsymbol{b}_j\, \boldsymbol{b}_j \ldots \boldsymbol{b}_j] \\ &= [\boldsymbol{b}_j\, \boldsymbol{Q}_j] \begin{bmatrix} 1 & 1 & \ldots & 1 \\ \boldsymbol{x}_{em}^{(1)} & \boldsymbol{x}_{em}^{(2)} & \ldots & \boldsymbol{x}_{em}^{(m_j)} \end{bmatrix} \end{aligned} \tag{2.15}$$

where $\boldsymbol{Q}_j$ is an $n \times n$ matrix and $\boldsymbol{b}_j$ is an $n \times 1$ column vector.

Let us define

$$\boldsymbol{C} = \left[\boldsymbol{x}_{os}^{(1)}\, \boldsymbol{x}_{os}^{(2)} \ldots \boldsymbol{x}_{os}^{(m_j)}\right]^T, \tag{2.16}$$

$$\boldsymbol{D} = \left[\widehat{\boldsymbol{f}}(\boldsymbol{x}_{em}^{(1)})\, \widehat{\boldsymbol{f}}(\boldsymbol{x}_{em}^{(2)}) \ldots \widehat{\boldsymbol{f}}(\boldsymbol{x}_{em}^{(m_j)})\right]^T = \begin{bmatrix} 1 & 1 & \ldots & 1 \\ \boldsymbol{x}_{em}^{(1)} & \boldsymbol{x}_{em}^{(2)} & \ldots & \boldsymbol{x}_{em}^{(m_j)} \end{bmatrix}^T \tag{2.17}$$

and

$$\boldsymbol{A}_j = \left[\boldsymbol{b}_j\, \boldsymbol{Q}_j\right]. \tag{2.18}$$

Then (2.14) becomes

$$\boldsymbol{C}^T = \boldsymbol{A}_j\, \boldsymbol{D}^T. \tag{2.19}$$

Transposing both sides of (2.19) gives

$$\boldsymbol{D}\, \boldsymbol{A}_j^T = \boldsymbol{C}. \tag{2.20}$$

Augmenting (2.20) by some weighting factors defined by an $m_j \times m_j$ diagonal matrix $\boldsymbol{W}$, where

$$\boldsymbol{W} = diag\{w_i\} \tag{2.21}$$

gives

$$\boldsymbol{W}\,\boldsymbol{D}\,\boldsymbol{A}_j^T = \boldsymbol{W}\,\boldsymbol{C}. \tag{2.22}$$

The least-squares solution to this system is

$$\boldsymbol{A}_j^T = \left(\boldsymbol{D}^T\,\boldsymbol{W}^T\,\boldsymbol{W}\,\boldsymbol{D}\right)^{-1}\boldsymbol{D}^T\,\boldsymbol{W}^T\,\boldsymbol{W}\,\boldsymbol{C}. \tag{2.23}$$

Larger/smaller weighting factors emphasize/deemphasize the influence of the corresponding base points on the SM transformation.

2.2. Implementation. We now present a straightforward implementation of the SM algorithm. First, begin with a point, $\boldsymbol{x}_{os}^* \triangleq arg\ min\{H(\boldsymbol{x}_{os})\}$, representing the optimal solution in $\boldsymbol{X}_{os}$ where $H(\boldsymbol{x}_{os})$ is some appropriate objective function. Then, the algorithm proceeds as follows:

Step 0. Initialize $\boldsymbol{x}_{em}^{(1)} = \boldsymbol{x}_{os}^*$. If $||\boldsymbol{R}_{os}(\boldsymbol{x}_{os}^*) - \boldsymbol{R}_{em}(\boldsymbol{x}_{em}^{(1)})|| \leq \epsilon$, stop. Otherwise, initialize $\Delta\boldsymbol{x}_{em}^{(i-1)}$ for $i = 2, 3, \ldots, m$.
Step 1. Select $m - 1$ additional base points in $\boldsymbol{X}_{em}$ by perturbation, $\boldsymbol{x}_{em}^{(i)} = \boldsymbol{x}_{em}^{(1)} + \Delta\boldsymbol{x}_{em}^{(i-1)}$ for $i = 2, 3, \ldots, m$.
Step 2. Perform m parameter extraction optimizations to obtain $\boldsymbol{x}_{os}^{(i)}$ for $i = 1, 2, \ldots, m$.
Step 3. Initialize $j = 0, m_j = m$.
Step 4. Compute $\boldsymbol{A}_j^T = \left(\boldsymbol{D}^T\,\boldsymbol{W}^T\,\boldsymbol{W}\,\boldsymbol{D}\right)^{-1}\boldsymbol{D}^T\,\boldsymbol{W}^T\,\boldsymbol{W}\,\boldsymbol{C}$ and extract the matrix $\boldsymbol{Q}_j$ and the vector $\boldsymbol{b}_j$ according to $\boldsymbol{A}_j = \left[\boldsymbol{b}_j\ \boldsymbol{Q}_j\right]$.
Step 5. Set $\boldsymbol{x}_{em}^{(m_j+1)} = \boldsymbol{Q}_j^{-1}(\boldsymbol{x}_{os}^* - \boldsymbol{b}_j)$.
Step 6. If $||\boldsymbol{R}_{os}(\boldsymbol{x}_{os}^*) - \boldsymbol{R}_{em}(\boldsymbol{x}_{em}^{(m_j+1)})|| \leq \epsilon$, stop.
Step 7. Perform parameter extraction optimization to obtain $\boldsymbol{x}_{os}^{(m_j+1)}$.
Step 8. Augment the matrix $\boldsymbol{C}$ with $\boldsymbol{x}_{os}^{(m_j+1)}$ and the matrix $\boldsymbol{D}$ with $\boldsymbol{x}_{em}^{(m_j+1)}$.
Step 9. Set $j = j + 1, m_j = m_j + 1$; go to *Step 4*.

Comments.

Note, that in *Steps 2* and *7* an auxiliary optimization (parameter extraction) is invoked. In *Step 5*, $\boldsymbol{x}_{em}^{(m_j+1)}$ may be snapped to the closest grid point if the EM simulator uses a fixed-grid meshing scheme.

2.3. Aggressive strategy for space mapping. An aggressive strategy has been derived for SM [7,8]. It exploits quasi-Newton iterations and Broyden update [12]. The aggressive approach is significantly more efficient than that described in the preceding subsection since it avoids performing time-consuming and possibly unproductive EM analyses at the perturbed points around the starting point. Instead, right from the beginning, it attempts to improve the solution in a systematic manner by well-targeted and fewer EM simulations. Moreover, the aggressive approach is better suited for automation of the SM technique [2].

3. Design of a high-temperature superconducting (HTS) parallel coupled-line microstrip filter exploiting Aggressive Space Mapping. The SM technique has been successfully applied to the design of an HTS filter. A detailed description of the filter (see Fig. 2) can be found in [3,6,7,8]. The filter uses microstrip technology where metallization strips are deposited on a solid dielectric substrate which separates the top metal strips and the bottom metal ground plane. Fig. 2 shows the top view of the metallization pattern. The dielectric and the ground plane box and their heights are not shown. The filter works as a two-port with the input and output ports located at the ends of leftmost and the rightmost strips. An electrical signal is transmitted from the input to the output by means of the electromagnetic field in the structure, including couplings between the strips. L_1, L_2 and L_3 are the lengths of the parallel coupled-line sections and S_1, S_2 and S_3 are the gaps between the sections. The width W is the same for all the sections as well as for the input and output microstrip lines, of length L_0. The thickness of the lanthanum aluminate substrate used is 20 mil. The dielectric constant and the loss tangent are assumed to be 23.425 and 3×10^{-5}, respectively. The metallization is considered lossless.

This relatively small circuit exemplifies difficulties in directly using detailed EM simulations during optimization. To obtain an accurate and detailed circuit response, such as the one drawn using the dashed line in Fig. 4, one needs more than a week of CPU time on a Sun SPARCstation 10. To invoke such simulations many times during optimization is prohibitive.

For this problem, we consider 6 optimization variables representing the geometrical dimensions of the filter: L_1, L_2, L_3, S_1, S_2 and S_3. We employ two models: (1) a fast model, based on empirical formulas available in the OSA90/hope [18] software package, and (2) an accurate but extremely CPU- intensive model, based on solving electromagnetic field equations by the ***em*** simulator [14].

Following *Phase 1* of SM, we optimize the HTS filter using the OSA90/hope empirical model. The optimization goal is formulated in terms of the so-called scattering parameters S. These S parameters quantify the filter behaviour in terms of the power transfer from the input to the output of the filter [17]. Of particular interest is the parameter $|S_{21}|$ and its dependence on frequency (frequency response). The filtering capabilities of the circuit considered are described by the design specifications:

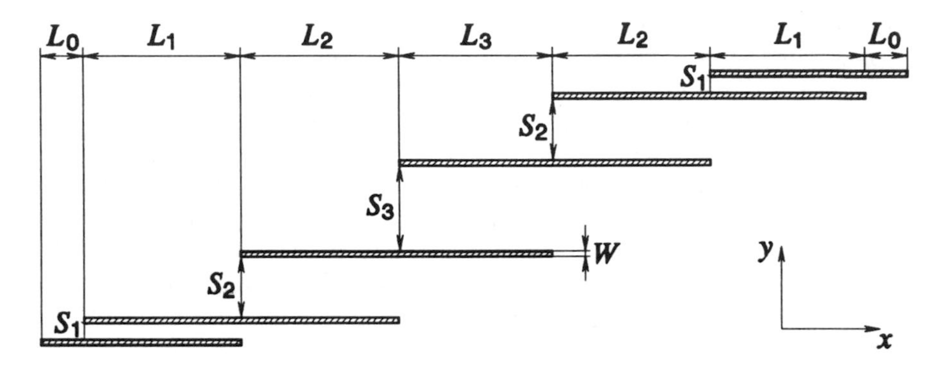

FIG. 2. *The structure of the HTS filter.*

$$|S_{21}| \leq 0.05 \quad \text{in the stopband}$$

$$|S_{21}| \geq 0.95 \quad \text{in the passband}$$

where the stopband includes frequencies below 3.967 GHz and above 4.099 GHz and the passband lies in the frequency range [4.008 GHz, 4.058 GHz]. An appropriate objective function for optimization is formulated from the design specifications [9]. Fig. 3 shows the $|S_{21}|$ (and $|S_{11}|$) empirical (fast) model responses after performing minimax optimization using OSA90/hope. Using the parameter values determined by minimax optimization the ***em*** simulated frequency response differs significantly from that of the empirical model, as shown in Fig. 4.

In *Phase 2* of SM, our aim is to establish a mapping in order to find a solution in the $\boldsymbol{X}_{em}$ space which substantially reproduces the performance predicted by the optimal empirical model. We report results obtained using the aggressive strategy (Section 2.3). The original approach (Sections 2.1 and 2.2) was also used and similar results were obtained. In this phase, in order to further reduce the CPU time of EM simulations, we do not consider as fine frequency sweeps as those shown in the figures. We use only 15 frequency points per sweep which turned out to be adequate. The SM solution emerges after only six such simplified EM analyses (13 for original SM). Fig. 5 compares the filter responses of the optimal empirical model and the ***em*** simulated SM solution.

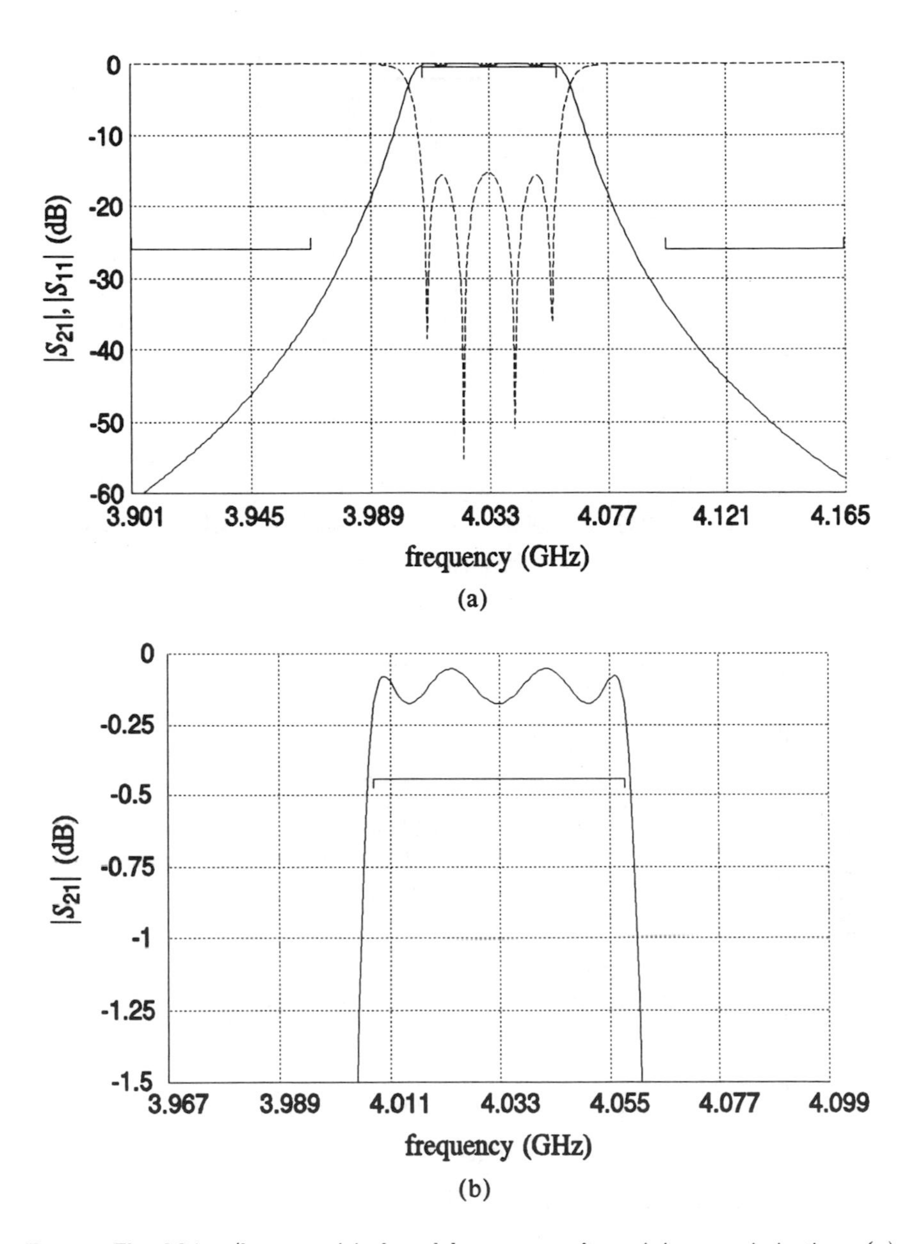

FIG. 3. *The OSA90/hope empirical model responses after minimax optimization. (a) $|S_{21}|$ (—) and $|S_{11}|$ (---) for the overall frequency band and (b) the passband details of $|S_{21}|$. ⊔ and ⊓ denote the upper and lower design specifications, respectively.*

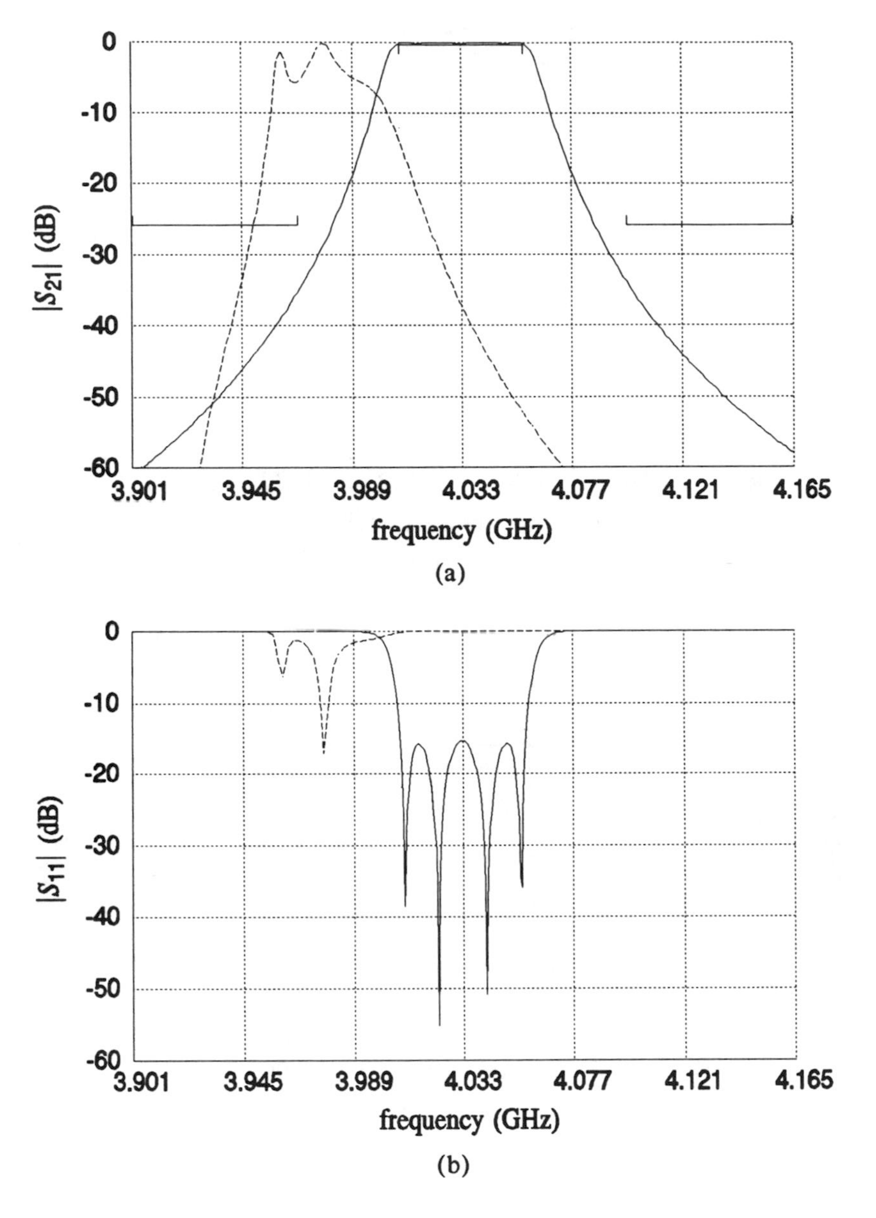

FIG. 4. *A comparison of (a)* $|S_{21}|$ *and (b)* $|S_{11}|$ *between the empirical model (—) and* **em** *(---) at the empirical model minimax solution.*

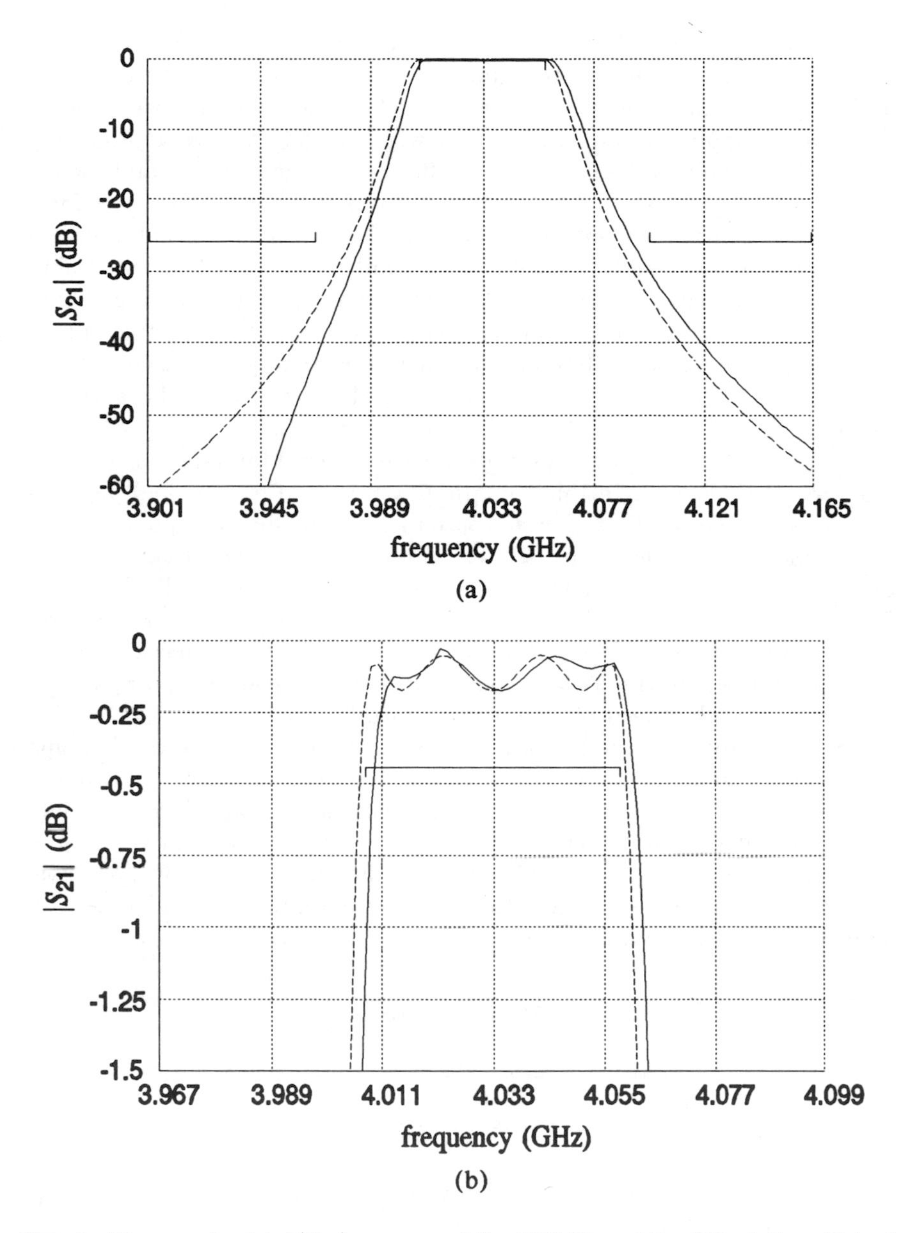

FIG. 5. *The* **em** *simulated* $|S_{21}|$ *response of the HTS filter at the SM solution obtained using the aggressive SM approach (—). The OSA90/hope empirical model solution (---) is shown for comparison. Responses are shown for (a) the overall frequency band and (b) the passband in more detail.*

4. Conclusions. This presentation has included a theoretical formulation of the Space Mapping technique and its application to the design of a high-temperature superconducting parallel coupled-line microstrip filter.

Space Mapping optimization is a newly emerging and very promising approach. It exploits the speed of an efficient surrogate model and blends it with a few slow evaluations of an accurate model to effectively perform design optimization within a practical time frame. A few recent publications in the engineering field exhibit some similarities to the concepts found in Space Mapping [13,15,16,21,22,23]. The main difference distinguishing Space Mapping from those techniques is that instead of trying to improve the surrogate models, it allows the input parameters of the fast model to be different from those in the actual physical object, or in the accurate model.

In the near future we expect to see the Space Mapping concept applied to active devices. In this domain, physics-based models and physical models will be utilized [1]. Physics-based models relate the equivalent circuit elements to the device physics based on simplified analytical solutions of device equations. Physical models, based on the numerical solution of fundamental device equations are the most accurate. However, they require significantly more computation time than the physics-based models. Hence, Space Mapping may be the key to achieving the accuracy of physical simulation and the speed of circuit-level optimization.

Surprisingly, the concept of Space Mapping is only now establishing itself in the domain of circuit optimization. This is despite the overwhelming array of engineering models of devices, circuits and systems. It should be noted that designers increasingly employ accurate CPU-intensive simulators, yet extensive use of efficient simplified models is made to avoid time-consuming analyses. SM bridges the two approaches and takes advantage of their respective benefits.

REFERENCES

[1] J.W. Bandler, *Statistical modeling, design centering, yield optimization and cost-driven design*, *CAD Design Methodology for Commercial Applications*. Workshop WFFE (A.M. Pavio, Organizer and Chairman), IEEE MTT-S Int. Microwave Symp. (Orlando, FL), 1995.

[2] J.W. Bandler, R.M. Biernacki, and S.H. Chen, *Fully automated space mapping optimization of 3D structures*, *IEEE MTT-S Int. Microwave Symp. Dig.* (San Francisco, CA), 1996, pp. 753–756.

[3] J.W. Bandler, R.M. Biernacki, S.H. Chen, W.J. Getsinger, P.A. Grobelny, C. Moskowitz and S.H. Talisa, *Electromagnetic design of high-temperature superconducting microwave filters*, *Int. J. Microwave and Millimeter-Wave Computer-Aided Engineering*, vol. 5, 1995, pp. 331–343.

[4] J.W. Bandler, R.M. Biernacki, S.H. Chen, P.A. Grobelny and R.H. Hemmers, *Exploitation of coarse grid for electromagnetic optimization*, *IEEE MTT-S Int. Microwave Symp. Dig.* (San Diego, CA), 1994, pp. 381–384.

[5] J.W. Bandler, R.M. Biernacki, S.H. Chen, P.A. Grobelny and R.H. Hemmers, *Space mapping technique for electromagnetic optimization*, *IEEE Trans. Microwave Theory Tech.*, vol. 42, 1994, pp. 2536–2544.

[6] J.W. Bandler, R.M. Biernacki, S.H. Chen, P.A. Grobelny, C. Moskowitz and S.H. Talisa, *Electromagnetic design of high-temperature superconducting microwave filters*, *IEEE MTT-S Int. Microwave Symp. Dig.* (San Diego, CA), 1994, pp. 993–996.

[7] J.W. Bandler, R.M. Biernacki, S.H. Chen, R.H. Hemmers and K. Madsen, *Aggressive space mapping for electromagnetic design*, *IEEE MTT-S Int. Microwave Symp. Dig.* (Orlando, FL), 1995, pp. 1455–1458.

[8] J.W. Bandler, R.M. Biernacki, S.H. Chen, R.H. Hemmers and K. Madsen, *Electromagnetic optimization exploiting aggressive space mapping*, *IEEE Trans. Microwave Theory Tech.*, vol. 43, 1995, pp. 2874–2882.

[9] J.W. Bandler and S.H. Chen, *Circuit optimization: the state of the art*, *IEEE Trans. Microwave Theory Tech.*, vol. 36, 1988, pp. 424–443.

[10] R.M. Biernacki, J.W. Bandler, J. Song and Q.J. Zhang, *Efficient quadratic approximation for statistical design*, *IEEE Trans. Circuits and Systems*, vol. 36, 1989, pp. 1449–1454.

[11] R.M. Biernacki and M.A. Styblinski, *Efficient performance function interpolation scheme and its application to statistical circuit design*, *Int. J. Circuit Theory and Appl.*, vol. 19, 1991, pp. 403–422.

[12] C.G. Broyden, *A class of methods for solving nonlinear simultaneous equations*, *Math. of Comp.*, vol. 19, 1965, pp. 577–593.

[13] S. Burgee, A.A. Guinta, R. Narducci, L.T. Watson, B. Grossman and R.T. Haftka, *A coarse grained variable-complexity approach to MDO for HSCT design*, in *Parallel Processing for Scientific Computing*, D.H. Bailey, P.E. Bjørstad, J.R. Gilbert, M.V. Mascagni, R.S. Schreiber, H.D. Simon, V.J. Torczon and L.T. Watson (eds.), SIAM, Philadelphia, PA, 1995, pp. 96–101.

[14] ***em***™ and ***xgeom***™, Sonnet Software, Inc., 1020 Seventh North Street, Suite 210, Liverpool, NY 13088.

[15] A.A. Giunta, V. Balabanov, M. Kaufman, S. Burgee, B. Grossman, R.T. Haftka, W.H Mason and L.T. Watson, *Variable-complexity response surface design of an HSCT configuration*, *Proceedings of the ICASE/LaRC Workshop on Multidisciplinary Design Optimization* (Hampton, VA), 1995.

[16] M.G. Hutchison, X. Huang, W.H. Mason, R.T. Haftka and B. Grossman, *Variable-complexity aerodynamic-structural design of a high-speed civil transport wing*, 4th AIAA/USAF/NASA/OAI *Symp. Multidisciplinary Analysis and Optimization* (Cleveland, OH), 1992, paper 92-4695.

[17] D.M. Pozar, *Microwave Engineering*, Addison-Wesley, 1990.

[18] *OSA90/hope*™ and *Empipe*™, Optimization Systems Associates Inc., P.O. Box 8083, Dundas, Ontario, Canada L9H 5E7.

[19] J. Sacks, W.J. Welch, T.J. Mitchell and H.P. Wynn, *Design and analysis of computer experiments*, *Statistical Science* vol. 4, 1989, pp. 409–435.

[20] D.G. Swanson, Jr., *Using a microstrip bandpass filter to compare different circuit analysis techniques*, *Int. J. Microwave and Millimeter-Wave Computer-Aided Engineering*, vol. 5, 1995, pp. 4–12.

[21] V.V. Toropov, *Simulation approach to structural optimization*, *Structural Optimization*, vol. 1, 1989, pp. 37–46.

[22] V.V. Toropov, A.A. Filatov and A.A. Polynkin, *Multiparameter structural optimization using FEM and multipoint explicit approximations*, *Structural Optimization*, vol. 6, 1993, pp. 7–14.

[23] S. Yesilyurt and A.T. Patera, *Surrogates for numerical simulations: optimization of eddy-promoter heat exchangers*, *Computer Methods Appl. Mech. Eng.*, vol. 121, 1995, pp. 231–257.

AN INVERSE PROBLEM IN PLASMA PHYSICS: THE IDENTIFICATION OF THE CURRENT DENSITY PROFILE IN A TOKAMAK

J. BLUM* AND H. BUVAT*

Abstract. This paper deals with the numerical identification of the plasma current density in a Tokamak, from experimental measurements. This problem consists in the identification of a non-linearity in a semi-linear 2D elliptic equation from Cauchy boundary measurements and from integrals of the magnetic field over several chords. A simplified problem, corresponding to the cylindrical approximation, is first solved by a SQP algorithm, where the non-linearity is decomposed in a basis of B-splines and where Tikhonov regularization is used, the regularizing parameter being determined by a cross-validation procedure. An alternative approach based on the decomposition of the non-linearity in a basis of compactly supported wavelets and the use of a new scalar product in the H^2 space has enabled to solve the identification problem in a robust way, without using Tikhonov regularization. Finally several test-cases are presented for the determination of the current density in the real toroidal configuration from magnetic, interferometric and polarimetric measurements.

1. Introduction. The reaction of nuclear fusion is the source of the radiation energy emitted by the stars: these reactions take place under gigantic pressures (10^{11} atm) due to gravitational forces and the masses of the stars.

To control thermonuclear fusion in the laboratory, the simplest reaction envisaged for the first fusion reactors is the Deuterium-Tritium reaction:

$$ {}_1^2D + {}_1^3T \rightarrow {}_2^4He + {}_0^1n + 17.6 \text{ Mev}. $$

The energy balance for such a "thermonuclear" environment should be positive, that is to say the energy liberated by the nuclear reaction ought to be greater than the thermal energy losses of the ionized gas called the plasma. This condition, known as Lawson's criterion, can be written:

$$ N\tau > L(T) $$

where N is the charged particle density, τ is the energy confinement time, T is the temperature of the plasma, and where the function L represents the limit of the ignition domain. For a temperature of 10 Kev ($\simeq$ 100 million degrees), $N\tau$ has to be at least equal to 10^{20} $m^{-3} \times s$. Research in thermonuclear fusion is aimed at realizing an experimental device which permits confinement of the plasma while satisfying Lawson's criterion.

Two approaches are possible. One aims to confine the plasma for a very short time τ but at very high density N: this is fusion by inertial confinement, where laser beams (or beams of electrons or ions) converge on a

* IDOPT Project (CNRS-INPG-INRIA-UJF), Tour IRMA, LMC/IMAG, BP 53, 38041 Grenoble Cedex 9, France.

target (plasma) in order to bring it to the thermonuclear state. The second approach is that of magnetic confinement where the ionized particles are confined within a magnetic field. A charged particle essentially describes a helix centered on the field line, and in order to confine the particle it suffices to maintain this field line on a closed surface, which necessarily must have a toroidal configuration. The density is much lower than in inertial confinement and is of the order of $10^{20}\ m^{-3}$; the confinement time must then be more significant (of the order of a second). The Tokamak device corresponds to this second method.

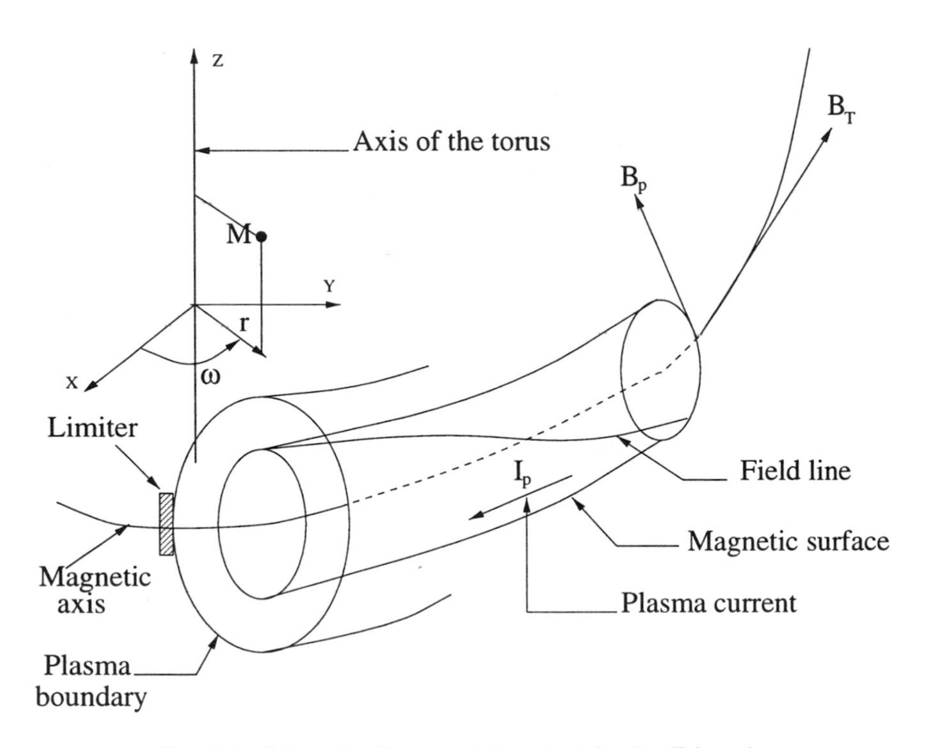

FIG. 1.1. *Schematic diagram of the principle of a Tokamak.*

Figure 1.1 represents a schematic diagram of the Tokamak principle. The magnetic field in the plasma region is the resultant of a poloidal field B_P (in the cross-section of the torus), generated by the plasma current I_P and by external coils, and a toroidal field B_T produced by coils wound around the torus. This poloidal field is required by the fact that, in order to have an equilibrium, the plasma pressure should be balanced by magnetic forces. The field lines generate magnetic surfaces which have the topology of nested tori. The magnetic axis corresponds to the case where the magnetic surface degenerates into a closed curve. The boundary of the plasma is a particular magnetic surface, defined by its tangency with a limiter, which prevents the plasma from touching the vacuum vessel. The plasma current I_P is obtained by induction from currents in poloidal field coils; the plasma thus appears as the secondary of a transformer whose poloidal field coils constitute the primary.

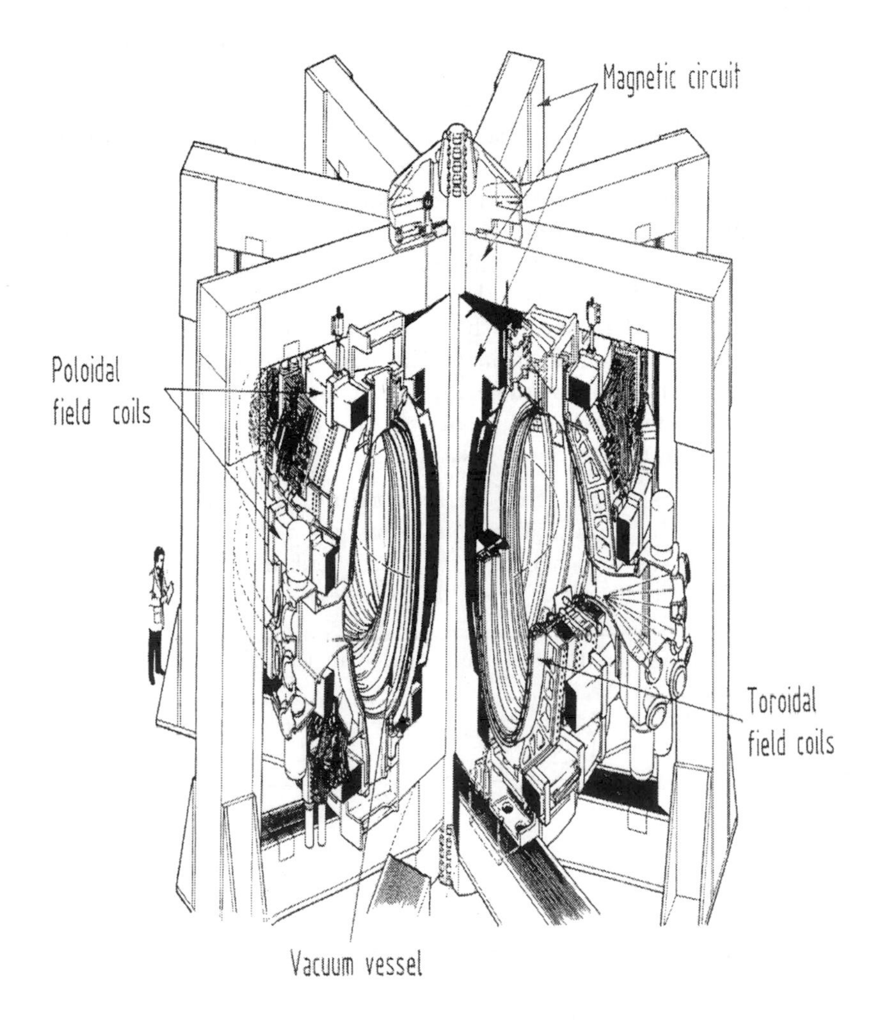

FIG. 1.2. *European Tokamak JET. (Photo: JET JOINT UNDERTAKING)*

Figure 1.2 represents the European Tokamak JET, where a magnetic circuit couples the primary poloidal field coils and the plasma. The vacuum vessel which contains the plasma, has a D-shaped cross-section.

It is of great importance to determine the profile of current density in a Tokamak for several reasons:

- the ohmic heating source is proportional to the square of this current density,
- the shape of the flux lines, and hence of the toroidal magnetic surfaces generated by the rotation of the flux lines around the z-axis, is related to the plasma current profile,
- the transport of particle and energy is anisotropic and the tensor of transport coefficients is linked to the geometry of the flux lines; the electronic and ionic densities and temperatures are approximately constant on each flux line and their diffusion is therefore perpendicular to the flux lines.

The aim of this paper is to identify the plasma current density from external experimental measurements, such as magnetic measurements, which are located on the vacuum vessel, or interferometric and polarimetric measurements inside the plasma. Mathematically the problem consists in identifying a non-linearity in an elliptic p.d.e. from Cauchy boundary measurements and from integrals over several chords.

Section 2 is devoted to the mathematical modelling of the problem. Section 3 concerns the identification of the non-linearity in a simplified case, where a cylindrical approximation is made for the Tokamak. Finally section 4 gives the numerical results obtained for the identification of the current density in the real toroidal Tokamak configuration.

2. Mathematical modelling of axisymmetric equilibrium for the plasma in a Tokamak. The equations which govern the equilibrium of a plasma in the presence of a magnetic field are on the one hand Maxwell's equations and on the other hand the equilibrium equations for the plasma itself.

Maxwell's equations as follows are satisfied in the whole of space (including the plasma):

$$\left\{ \begin{array}{rcl} \nabla . B & = & 0 \\ \nabla \times H & = & j \\ B & = & \mu H \end{array} \right. \tag{2.1}$$

where B and H represent the induction and the magnetic field respectively, μ is the magnetic permeability and j is the current density. The first relation of (2.1) is the equation of conservation of magnetic induction, the second one is Ampere's Theorem, and the third one is linear in vacuum and plasma, where $\mu = \mu_0$.

The equilibrium equation for the plasma is

$$\nabla p = j \times B. \tag{2.2}$$

This equation (2.2) signifies that the plasma is in equilibrium when the force ∇p due to the kinetic pressure p is equal to the force of the magnetic pressure $j \times B$. We deduce immediately from (2.2) that

$$B . \nabla p = 0 \tag{2.3}$$

$$j.\nabla p = 0. \tag{2.4}$$

Thus in a plasma in equilibrium the field lines and the current lines lie on isobaric surfaces ($p =$ const.); these surfaces, generated by the field lines, are called magnetic surfaces. In order that they should remain within a bounded volume of space it is necessary that they have toroidal topology. Each surface $p = C$ (except for a set of values of C of measure zero) is traversed ergodically by the flux lines, and we suppose that these surfaces form a family of nested tori. The innermost torus degenerates into a curve which we call the magnetic axis.

In the cylindrical coordinate system (r, z, ω) of figure 1.1, the hypothesis of axial symmetry consists in supposing that the magnetic induction B is independent of the toroidal angle ω. From equation (2.1), one can then define a poloidal flux $\Psi(r, z)$ such that:

$$\left\{\begin{array}{rcl} B_r & = & -\dfrac{1}{r}\dfrac{\partial\Psi}{\partial z} \\ B_z & = & \dfrac{1}{r}\dfrac{\partial\Psi}{\partial r} \end{array}\right. \tag{2.5}$$

As far as the toroidal component B_T of the induction B is concerned, we define F by

$$B_T = \frac{F}{r}e_T \tag{2.6}$$

where e_T is the unit vector in the toroidal sense. Then the magnetic induction B can be written as:

$$\left\{\begin{array}{rcl} B & = & B_P + B_T \\ B_P & = & \dfrac{1}{r}[\nabla\Psi \times e_T] \\ B_T & = & \dfrac{F}{r}e_T \end{array}\right. \tag{2.7}$$

where B_P denotes the poloidal component of B. According to (2.7), in an axisymmetric configuration the magnetic surfaces are generated by the rotation of the flux lines $\Psi =$ constant around the vertical axis Oz of the torus.

From (2.7) and the last two relations of (2.1) we obtain the following expression for j:

$$\left\{\begin{array}{rcl} j & = & j_P + j_T \\ j_P & = & \dfrac{1}{r}\left[\nabla\left[\dfrac{F}{\mu}\right] \times e_T\right] \\ j_T & = & (-\Delta^*\Psi)e_T \end{array}\right. \tag{2.8}$$

where j_P and j_T are the poloidal and toroidal components respectively of j, and the operator Δ^* is defined by

$$\Delta^*. = \frac{\partial}{\partial r}\left[\frac{1}{\mu r}\frac{\partial .}{\partial r}\right] + \frac{\partial}{\partial z}\left[\frac{1}{\mu r}\frac{\partial .}{\partial z}\right]. \tag{2.9}$$

The expressions (2.7) and (2.8) for B and j are valid in the whole of space since they involve only Maxwell's equations and the hypothesis of axisymmetry.

If we consider now the plasma region, the relation (2.3) implies that ∇p is collinear with $\nabla\Psi$, and therefore p is constant on each magnetic surface: we can denote this by

$$p = p(\Psi). \tag{2.10}$$

Relation (2.4) combined with the expression (2.8) for j implies that ∇F is collinear with ∇p, and therefore that F is likewise constant on each magnetic surface:

$$F = F(\Psi). \tag{2.11}$$

The equilibrium relation (2.2) combined with the expressions (2.7) and (2.8) for B and j implies that:

$$\nabla p = -\frac{\Delta^*\Psi}{r}\nabla\Psi - \frac{F}{\mu_0 r^2}\nabla F. \tag{2.12}$$

If we use the notation

$$\frac{\nabla p}{\nabla\Psi} = \frac{\partial p}{\partial\Psi}, \qquad \frac{\nabla F}{\nabla\Psi} = \frac{\partial F}{\partial\Psi},$$

then (2.12) can be written:

$$-\Delta^*\Psi = r\frac{\partial p}{\partial\Psi} + \frac{1}{2\mu_0 r}\frac{\partial F^2}{\partial\Psi}. \tag{2.13}$$

This equation (2.13) is called the Grad-Shafranov equilibrium equation. The operator Δ^* is an elliptic linear operator since μ is equal to μ_0 in the plasma. By (2.8), the right-hand side of (2.13) represents the toroidal component of the plasma current density. It involves the functions p and F which are not directly measured inside the plasma.

The aim of the following sections of this paper will be to identify the right-hand side of equation (2.13), i.e. the non-linearity of this elliptic equation.

3. Numerical identification of the non-linearity in the cylindrical case. Let us consider first a simplified case, corresponding to the cylindrical case, which is the limit of the toroidal one, when the aspect ratio of the torus goes to infinity. In this case, Grad-Shafranov equation can be simplified and the magnetic flux Ψ is related to the current density $f \geq 0$ by the following semilinear elliptic equation:

$$\left\{ \begin{array}{rcl} -\Delta\Psi & = & f(\overline{\Psi}) \text{ in } \Omega, \\ \Psi & = & 0 \text{ on } \Gamma = \partial\Omega; \end{array} \right. \tag{3.1}$$

with $\overline{\Psi} = \dfrac{\Psi}{\max\limits_{\Omega} \Psi} \in [0,1]$ from the maximum principle (this normalized flux is introduced so that f will be identified on the fixed interval [0,1]).

The magnetic field $\partial\Psi/\partial n$ along the plasma boundary Γ is assumed to be measured experimentally; let us call g this function. The identifiability of f from g depends on the shape of Γ. Indeed, if Γ is a circle, assuming f C^1 and positive, from [1] the solution Ψ of (3.1) is radial and hence $\partial\Psi/\partial n$ is constant on Γ. From this constant g, it is impossible to identify f in a unique way as there exists an infinite number of functions f giving the same g (provided that $\int\limits_{\Omega} f dx = -g \times |\Gamma|$, where $|\Gamma|$ is the length of Γ). The degeneracy of this case is linked to the fact that $\partial\Psi/\partial n$ is constant on Γ and it has been proven in [2] that the only C^2 domain where this happens is a disk. Hence one might conjecture that, for smooth positive functions f, it is possible to identify f in a unique way from g, if the domain Γ is non-circular. This problem has been solved in [3] for domains with corners (X-points in the terminology of plasma physicists), where the uniqueness of the determination of positive analytic functions f from g is proved.

This problem of identifiability remains open for smooth non-circular domains. Partial results have been obtained in [4], for the case where f is an affine function, and where Ω is a bounded strictly convex $C^{3,\alpha}$ domain in $\mathbb{R}^2$, which is not a disk; it has been proven that there exists at most a finite number of affine functions f that correspond to a given g.

The numerical identification problem is formulated as a "least-square" minimization with a "Tikhonov regularization", the cost-function J_ε being defined as:

$$J_\varepsilon(f) = \int\limits_{\Gamma} \left(\frac{\partial\Psi}{\partial n} - g \right)^2 d\sigma + \varepsilon \int\limits_0^1 f''(x)^2 dx; \tag{3.2}$$

where Ψ is related to f by (3.1) and where $f \in \mathcal{U}_{\text{ad}}$, $\mathcal{U}_{\text{ad}}$ being the set of C^2-functions defined on the interval [0,1] such that $f(0) = 0$. The second term of the cost-function is required by the ill-posedness of this inverse problem, which consists in the determination of f from Cauchy boundary measurements and it enables to obtain smooth functions for which the inverse problem becomes well-posed.

A similar least-square minimization (without regularization) has been used in [5] in order to identify third order polynomials for f (in the case of exact data).

3.1. The Lagrangian and the optimality system. Since the system of state equations is non-linear, and the cost-function J_ε non convex, we shall define a Lagrangian in order to derive formally the conditions for the optimization of (3.2):

$$\mathcal{L}_\varepsilon(f, \Psi, p) = J_\varepsilon(f, \Psi) + \int_\Omega \nabla\Psi . \nabla p dx - \int_\Omega f(\overline{\Psi}) p dx$$

A sufficient condition for (f, Ψ) to be a minimizer to the functional (3.2) and p to be the associated Lagrange multiplier is that (f, Ψ, p) should be a saddle point for $\mathcal{L}_\varepsilon$.

Under the assumptions:
$(H1)$ $\max_\Omega \Psi$ is attained at one and only one point M_0 in the interior of Ω,
$(H2)$ Ψ is of class $\mathcal{C}^2$ in a neighbourhood of M_0 and the point M_0 is a non degenerate elliptic point, we can prove that $b : \Psi \mapsto \overline{\Psi} = \dfrac{\Psi}{\max_\Omega \Psi}$ is Gâteaux differentiable and its derivative $b'(\Psi)$ is the map which associates $\dfrac{\Psi(M_0)\Phi - \Psi\Phi(M_0)}{\Psi(M_0)^2}$ to Φ (see [6]). Hence, if $f \mapsto \Psi$ satisfying (3.1) is Gâteaux differentiable:

$$\widetilde{\Psi} = \lim_{\theta \to 0} \frac{\Psi(f + \theta \widetilde{f}) - \Psi(f)}{\theta}$$

then the derivative $\widetilde{\Psi}$ satisfies:

$$\int_\Omega \nabla\widetilde{\Psi} . \nabla\Phi dx - \int_\Omega f'(\overline{\Psi}) \left(b'(\Psi)\widetilde{\Psi}\right) \Phi dx = \int_\Omega \widetilde{f}(\overline{\Psi})\Phi dx, \ \forall\Phi. \tag{3.3}$$

If hypotheses $(H1)$ and $(H2)$ are satisfied, then we can deduce the optimality system from the Euler-Lagrange conditions:

$$\begin{cases} \left(\dfrac{\partial \mathcal{L}_\varepsilon}{\partial \Psi}, \Phi\right) & = \ 0, \ \forall\Phi, \\ \left(\dfrac{\partial \mathcal{L}_\varepsilon}{\partial f}, \phi\right) & = \ 0, \ \forall\phi \in \mathcal{U}_{\mathrm{ad}}, \end{cases}$$

from which we have the weak formulation for the adjoint state p and the gradient of the cost-function (for (f, Ψ) satisfying the state equations (3.1)):

$$\int_\Omega \nabla\Phi . \nabla p dx - \int_\Omega f'(\overline{\Psi})(b'(\Psi)\Phi) p dx = \int_\Gamma \left(g - \frac{\partial \Psi}{\partial n}\right) \frac{\partial \Phi}{\partial n} d\sigma, \ \forall\Phi \tag{3.4}$$

$$(3.5) \qquad (\nabla J_\varepsilon(f), \phi) = \varepsilon \int_0^1 f''(x)\phi''(x)dx - \int_\Omega \phi(\overline{\Psi})pdx, \ \forall \phi \in \mathcal{U}_{\text{ad}}.$$

3.2. Identification in a basis of cubic B-splines. The function f is decomposed in a basis of cubic B-splines $N_i^4(x)$:

$$f(\overline{\Psi}) = \sum_{i=1}^{M} a_i N_i^4(\overline{\Psi}).$$

These functions $N_i^4(x)$ are piecewise polynomials of degree 3, positive for all x, compactly supported and of class $\mathcal{C}^2([0,1])$. Hence, $J_\varepsilon(f)$ can be written as $J_\varepsilon(a_1, \ldots, a_M)$ and its minimization is achieved by a sequential quadratic programming (SQP) method.

3.2.1. The SQP method. The SQP method consists of external iterations, each one being the solution of a linear quadratic control problem and consisting of a sequence of internal conjugate gradient iterations. More precisely, let V_h be the discretized state space, then the nth external iteration of the algorithm consists of defining for the linearized problem (3.3) in $(f^n, \Psi^n) \in \mathbb{R}^M \times V_h$ a control problem (P^n) with a quadratic functional J_ε^n "close to" J_ε. This linear quadratic control problem will itself be solved by a succession of internal iterations of conjugate gradient, and (f^{n+1}, Ψ^{n+1}) will thus be computed as being the optimum for the problem (P^n).

• **external iterations:**

⋆ let $(a^0, \Psi^0) \in \mathbb{R}^M \times V_h$,

⋆ n^{th} external iteration: the state equations are linearized with respect to Ψ and to the a_i, and become:

$$(3.6) \qquad \begin{aligned} \int_\Omega \nabla\Psi.\nabla\Phi dx \ &- \sum_{i=1}^{M} a_i^n \int_\Omega N_i^{4\prime}(\overline{\Psi^n})(b'(\Psi^n)\Psi)\Phi dx \\ &= \sum_{i=1}^{M} a_i \int_\Omega N_i^4(\overline{\Psi^n})\Phi dx, \ \forall \Phi \in V_h. \end{aligned}$$

The nth optimization problem (P^n) is then:

Find $(a^{n+1}, \Psi^{n+1}) \in \mathbb{R}^M \times V_h$ satisfying (3.6) and such that

$$J^n(a^{n+1}, \Psi^{n+1}) = \inf_{(a,\Psi) \in \mathbb{R}^M \times V_h} J^n(a, \Psi)$$

with $J^n(a, \Psi) = \displaystyle\int_\Gamma \left(\frac{\partial\Psi}{\partial n} - g\right)^2 d\sigma + \varepsilon a^t \Lambda a$, *$(a, \Psi)$ satisfying (3.6), and*

where Λ is the $M \times M$ matrix: $\Lambda_{ij} = \displaystyle\int_0^1 N_i^{4\prime\prime}(x)N_j^{4\prime\prime}(x)dx.$

The internal iterations correspond to the minimization of the cost function J^n by conjugate gradient method. The optimality system corresponding to the problem (P^n) is then:

$$(3.7)\quad \left\{\begin{aligned}
&\int_\Omega \nabla\Psi^{n+1}.\nabla\Phi dx - \sum_{i=1}^M a_i^n \int_\Omega N_i^{4\prime}(\overline{\Psi^n})(b'(\Psi^n)\Psi^{n+1})\Phi dx \\
&\qquad = \sum_{i=1}^M a_i^{n+1} \int_\Omega N_i^4(\overline{\Psi^n})\Phi dx, \\
&\int_\Omega \nabla\Phi.\nabla p^{n+1} dx - \sum_{i=1}^M a_i^n \int_\Omega N_i^{4\prime}(\overline{\Psi^n})(b'(\Psi^n)\Phi)p^{n+1} dx \\
&\qquad = \int_\Gamma \left(g - \frac{\partial\Psi^{n+1}}{\partial n}\right)\frac{\partial\Phi}{\partial n} d\sigma, \\
&\varepsilon(\Lambda a^{n+1})_i - \int_\Omega N_i^4(\overline{\Psi^n})p^{n+1} dx = 0,\ \forall i = 1,\ldots,M
\end{aligned}\right.$$

with $(a^{n+1}, \Psi^{n+1}, p^{n+1}) \in \mathbb{R}^M \times V_h \times V_h$, $\forall\Phi \in V_h$.

3.2.2. Choice of the regularization parameter ε. The choice of the regularization parameter ε is very important in our identification problem. A too small value of ε makes the problem ill-posed again, and a too big value of ε shall give a too smooth solution. In order to determine ε, we use a cross validation method which principle is to divide the data g in two subsets; the first in order to compute an estimation of the solution f for fixed ε, the second one to evaluate if the obtained value for ε is well chosen.

We consider now our model as:

$$z = T(a) + \nu,$$

z being the discretized data, $T(a) = \left(\dfrac{\partial\Psi}{\partial n}\right)_\Gamma$ discretized, and ν is the noise, which law is supposed gaussian $\mathcal{N}(0, \sigma^2)$.

We temporarily forget the kth data z_k, and we note $z_\varepsilon^{(k)} = T(a_\varepsilon^{(k)})$ where $a_\varepsilon^{(k)}$ is solution to the problem:

$$\min_{a\in\mathbb{R}^M} \sum_{\substack{i=1\\ i\neq k}}^{n} \alpha_i^2 (T(a)_i - z_i)^2 + \varepsilon a^t \Lambda a$$

where the α_i^2 are positive weights which are due to the discretization by finite elements.

The prediction error on the kth forgotten data is $z_{\varepsilon,k}^{(k)} - z_k$, hence we shall choose ε which minimizes the quadratic mean of these weighted errors for all k:

$$\mathcal{V}_0(\varepsilon) = \frac{1}{n}\sum_{k=1}^{n} \alpha_k^2 (z_{\varepsilon,k}^{(k)} - z_k)^2.$$

In order to minimize the cross validation function $\mathcal{V}_0(\varepsilon)$, we write $\varepsilon = 10^{-\gamma}$ with $\gamma \in [0, 20]$, and we choose for γ the minimum value for $\mathcal{V}_0(10^{-\gamma})$ obtained by a "golden section" algorithm. The numerical cost of this cross validation is high because it needs the resolution of n optimization problems, where n is the data number.

3.2.3. Numerical results. The boundary $\Gamma = \partial\Omega$ we consider here is an ellipse, and the sensitivity of the quality of the identification is studied in terms of the elongation of the ellipse.

A conformal mapping is used to transform the ellipse onto a circle and a linear finite element method enables to solve the equation for Ψ and p in the disk. Simulated data are introduced for the Neumann boundary condition g with an amount of noise of 5%. The number M of B-splines, constituting the basis of the approximation space for f, has been taken equal to 7.

Figures 3.1 and 3.2 present the results for various values of the elongation e. It appears that when the elongation becomes greater than 1.25, the identification of f is possible and correct. When the boundary becomes close to a circle, f cannot be identified, as it is expected from the theoretical results.

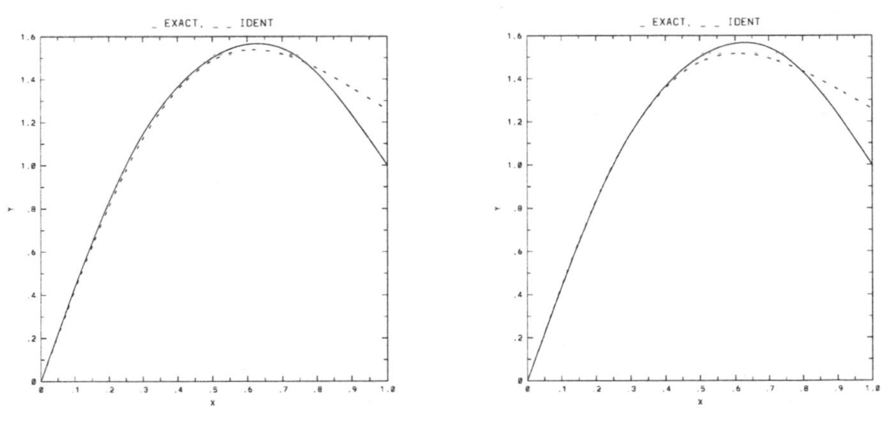

FIG. 3.1. *Left: e=2, right: e=1.5*

The influence of the regularization parameter is illustrated by the figure 3.3. The considered domain is delimited by an ellipse of elongation $e = 2$. The data are noised at 5%. It appears clearly that for ε greater than ε^*

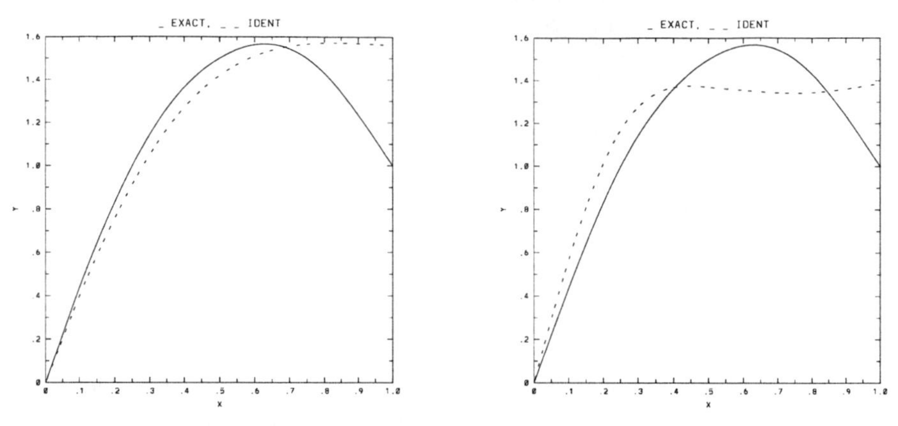

FIG. 3.2. *Left: e=1.2, right: e=1.01*

obtained by cross validation, the identified function f is too smooth, while instabilities appear for ε smaller than ε^*.

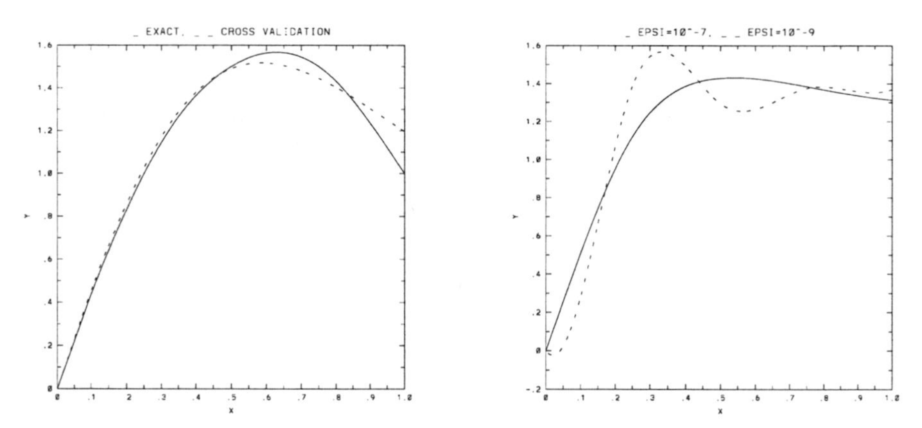

FIG. 3.3. *Left: identification with* $\varepsilon^* \simeq 10^{-8}$ *given by cross validation, right: identification with one value smaller than* ε^* *and one greater.*

3.3. Identification in a basis of scaling functions. The choice of the regularization parameter is very expensive. To avoid this problem, we consider a new basis of functions for the control space $\mathcal{U}_{\text{ad}}$. More precisely, we consider a multiresolution analysis (see [8]):

$$\ldots \subset V_{-1} \subset V_0 \subset V_1 \subset \ldots \subset L^2(\mathbb{R}), \quad L^2(\mathbb{R}) = \overline{\left(\bigcup_{j\in\mathbb{Z}} V_j\right)}, \quad \bigcap_{j\in\mathbb{Z}} V_j = \{0\}.$$

We consider a particular approximation space V_j (for fixed j) and the scaling functions

$$\{\phi_{j,k},\ k \in \mathbb{Z} \text{ such that } (Supp\ \phi_{j,k}) \bigcap [0,1] \neq \phi\}$$

associated to the Daubechies compactly supported wavelets (see [7]). The restrictions of these functions $\phi_{j,k}$, for $k \in \mathbb{Z}$, to the interval $[0,1]$ are no longer an orthonormal basis for V_j. Let $X = H^2_{\alpha,\beta,\gamma}(0,1)$ be the Hilbert space (with $\alpha > 0$, $\beta \geq 0$ and $\gamma \geq 0$) associated to the scalar product:

$$\langle u, v\rangle_X = \alpha \int_0^1 u(x)v(x)dx + \beta \int_0^1 u'(x)v'(x)dx + \gamma \int_0^1 u''(x)v''(x)dx.$$

We orthogonalize the basis $\{\phi_{j,k},\ k \in \mathbb{Z}\}$ (for fixed j) with respect to this scalar product, by computing the eigenvalues λ_k and eigenvectors w_k of the Gram matrix $G = (G_{\ell m})$ with $G_{\ell m} = \langle \phi_{j,\ell}, \phi_{j,m}\rangle_X$. We define then the new functions $\widetilde{\phi}_{j,i}(x) = \frac{1}{\lambda_i} \sum_{k \in \mathbb{Z}} (w_i)_k \phi_{j,k}(x)$ which form an orthonormal set for the scalar product of X and one searches for $f(\overline{\Psi}) = \sum_{i=1}^{M} b_i \widetilde{\phi}_{j,i}(\overline{\Psi})$ that minimizes J_0 (corresponding to $\varepsilon = 0$ in (3.2)), Ψ being related to f by (3.1). The same SQP method as in section 3.2.1 is used in order to minimize J_0 in this new basis, and M is taken equal to 20.

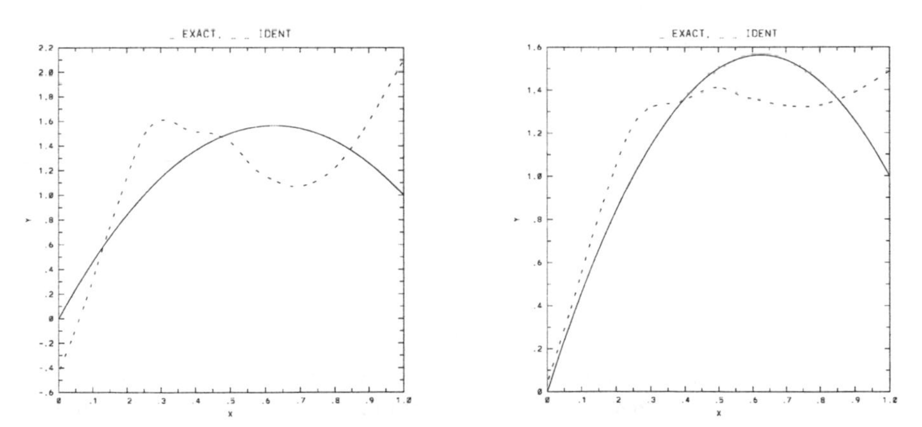

FIG. 3.4. *Identification in the orthonormal basis with respect to the scalar product of* $H^2_{1,1,\gamma}$*; left:* $\gamma = 1$*, right:* $\gamma = 10$.

Figures 3.4 and 3.5 present the results obtained for the same case as on figure 3.1 (left) with $\alpha = \beta = 1$ and different values of γ. It should be noted that, if $\gamma = 0$ the identification is unstable. If $X = H^2$ ($\gamma =$

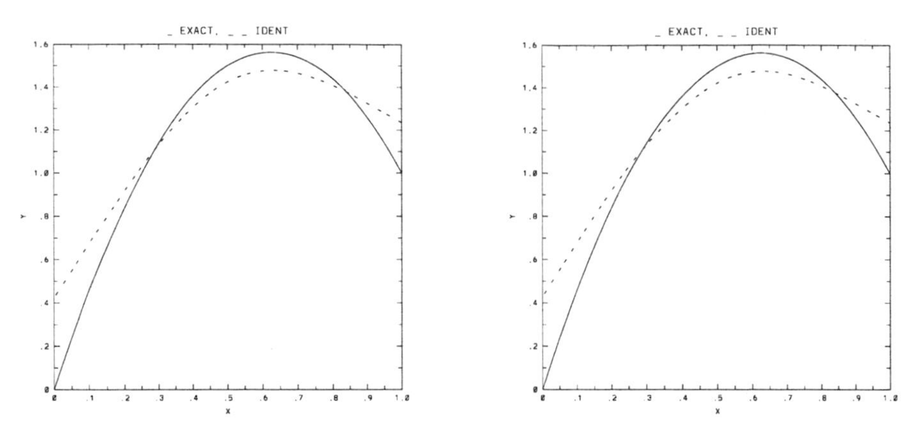

FIG. 3.5. *Identification in the orthonormal basis with respect to the scalar product of* $H^2_{1,1,\gamma}$*; left:* $\gamma = 100$*, right:* $\gamma = 1000$.

1), the identification is poor and a large weight γ on the second order derivatives is necessary in order to obtain a robust determination of f. Comparing the results on figure 3.1 with figure 3.5, it appears that the identification obtained in $H^2_{1,1,\gamma}$ with large γ is not as good as the one obtained with a Tikhonov regularization, but is less time consuming as there is no regularization parameter to be determined.
In conclusion, the determination of the non-linearity f is possible for non circular shapes and the use of the $H^2_{1,1,\gamma}$ spaces enables to identify f without any Tikhonov regularization.

4. Identification of the plasma current density from magnetic and polarimetric measurements. Figure 4.1 represents one eighty of the toroidal vacuum vessel, where are located the flux loops, measuring the poloidal flux Ψ, and the magnetic probes, measuring the tangential component of the magnetic field which is equal, according to (2.7), to $\frac{1}{r}\frac{\partial\Psi}{\partial n}$, where $\frac{\partial\Psi}{\partial n}$ is the derivative of Ψ normal to the vacuum vessel. Figure 4.2 represents the cross-section of the vacuum vessel with the locations of the flux loops on Γ_1 (external side of the vacuum vessel) and of the magnetic probes on Γ_2 (internal side). The vacuum vessel contains the plasma Ω_p and the vacuum region Ω_V. They are separated by the plasma boundary Γ_p, which is the particular flux line tangent to the limiter D, which prevents the plasma from touching the vacuum vessel. The plasma boundary is a free boundary, and hence a new non-linearity of the problem.

The equations, which govern the behaviour of $\Psi(r, z)$ in the domain

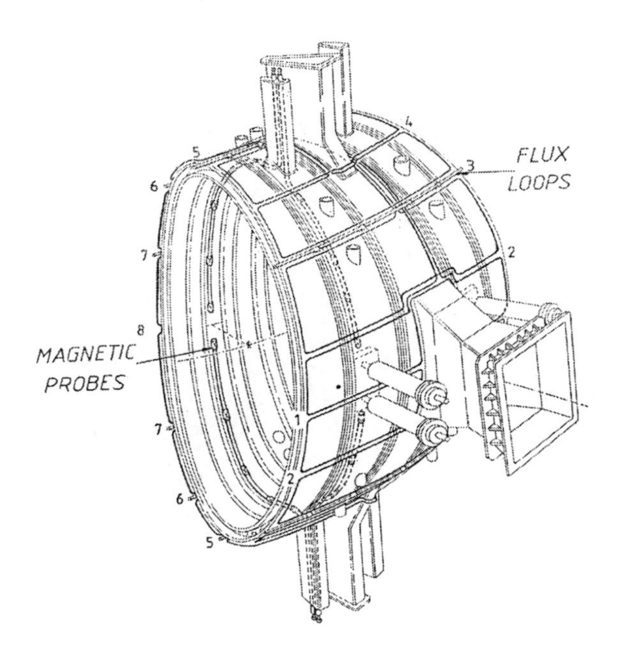

FIG. 4.1. *One eighty of the vacuum vessel*

limited by Γ_1, are:

$$(4.1)\qquad \begin{cases} -\Delta^* \Psi = [rA(\overline{\Psi}) + \dfrac{1}{r}B(\overline{\Psi})]\mathbb{1}_{\Omega_p} \text{ in } \Omega \\ \Psi = h \text{ on } \Gamma_1 \end{cases}$$

with $\Omega_p = \{x \in \Omega$ such that $\Psi(x) > \max_D \Psi\}$, where $A(\overline{\Psi}) = \dfrac{\partial p}{\partial \Psi}(\overline{\Psi})$, $B(\overline{\Psi}) = \dfrac{1}{2\mu_0}\dfrac{\partial F^2}{\partial \Psi}(\overline{\Psi})$, $\overline{\Psi} = \dfrac{\Psi - \max_{\Omega_p}\Psi}{\max_D \Psi - \max_{\Omega_p}\Psi} \in [0,1]$ in Ω_p, $\mathbb{1}_{\Omega_p}$ is the characteristic function of Ω_p and h is the Dirichlet boundary condition obtained by interpolation between the discrete measurements of Ψ at the points M_i of Γ_1. There is no current flowing in the vacuum region Ω_V, hence the r.h.s. of the equation for Ψ vanishes outside Ω_p, and is given by equation (2.13) inside Ω_p. The plasma boundary Γ_p is defined by:

$$\Gamma_p = \{x \in \Omega \text{ such that } \Psi(x) = \max_D \Psi\}$$

where D is the limiter. This condition defines the contact between the plasma and the limiter. Equation (4.1) is solved by a discretization method, using quadratic finite elements.

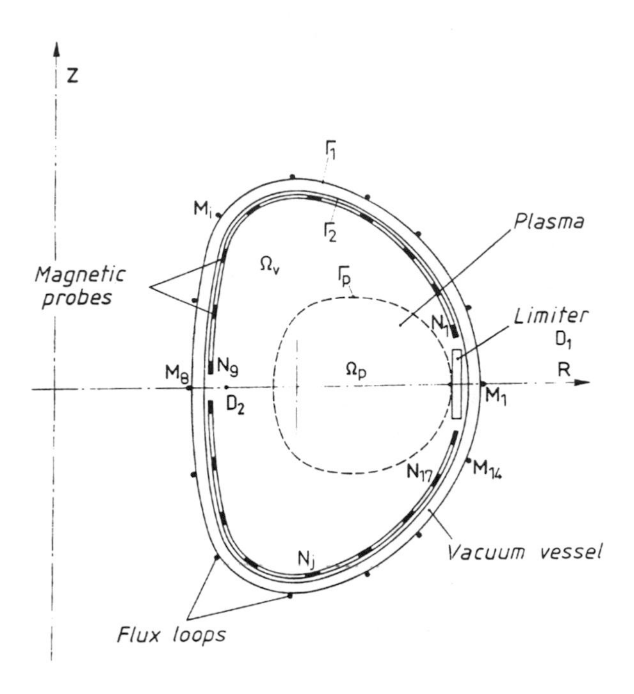

FIG. 4.2. *Cross-section of the vacuum vessel*

We are now going to identify, if possible, the two functions A and B in equation (4.1) from the experimental measurements which are:
• the magnetic measurements g of $\frac{1}{r}\frac{\partial \Psi}{\partial n}$ on the vacuum vessel,
• the polarimetric measurements that give the Faraday rotation of the angle of infrared radiation crossing the section of the plasma along several vertical chords C_i:

$$\alpha_i = \int_{C_i} n_e(\overline{\Psi}) B_z \, dz = \int_{C_i} \frac{n_e(\overline{\Psi})}{r} \frac{\partial \Psi}{\partial r} dz$$

where n_e represents the electronic density, which is approximately constant on each flux line,
• the interferometric measurements that give the density line integrals over the chords C_i:

$$N_i = \int_{C_i} n_e(\overline{\Psi}) d\ell.$$

As in section 3, the identification is formulated as a "least-square" minimization of:

$$J_\varepsilon(A,B,n_e) = \int_{\Gamma_2} \left(\frac{1}{r}\frac{\partial\Psi}{\partial n} - g\right)^2 d\sigma + K_1 \sum_{i=1}^{k} \left[\int_{C_i} \frac{n_e(\overline{\Psi})}{r}\frac{\partial\Psi}{\partial r} d\ell - \alpha_i\right]^2 + K_2 \sum_{i=1}^{k} \left[\int_{C_i} n_e(\overline{\Psi}) d\ell - N_i\right]^2 + \varepsilon_1 \int_0^1 [A''(x)]^2 dx + \varepsilon_2 \int_0^1 [B''(x)]^2 dx + \varepsilon_3 \int_0^1 [n_e''(x)]^2 dx \tag{4.2}$$

The weights K_1 and K_2 enable to give more or less importance to the corresponding experimental measurements ($K_1 = K_2 = 0$ corresponds to the case where there are only magnetic measurements), the coefficients ε_1, ε_2, ε_3 are the regularizing parameters. It should be noticed that the electronic density n_e does not intervene in system (4.1), but, as soon as we want to use the polarimetric measurements, it is necessary to include n_e (and hence interferometry) in the identification procedure.

The three functions A, B and n_e are decomposed in a basis of splines and their coefficients are then the optimization variables; the minimization of J_ε is performed as in section 3.2 by using a SQP algorithm.

Let us consider the following test-case:

$$A(x) = (1-x)^{\frac{3}{2}}, \quad B(x) = 1 - x^3, \quad n_e(x) = (1-x)^{\frac{1}{2}}.$$

Experimental data are simulated by solving system (4.1) and are then used as input in the identification procedure. A certain amount of noise is introduced on these pseudo experimental data, in order to test the robustness of the algorithm. Theoretically, it is possible to identify separately A and B from the magnetic measurements in domains with corners ([9]), but numerically it appears impossible to determine them separately from magnetic measurements only. We introduce then the polarimetric and interferometric measurements and the results of identification are given on figure 4.3 for $A(\overline{\Psi})$ and $B(\overline{\Psi})$. If we introduce the following average of j_T over a flux line L:

$$\langle j_T \rangle = \frac{\partial}{\partial V} \int_V j_T dV = \frac{\int_L \frac{j_T d\ell}{B_p}}{\int_L \frac{d\ell}{B_p}}$$

where V is the volume enclosed inside the magnetic surface generated by the flux line L, then figure 4.4 compares the real and the identified profile of $\langle j_T \rangle$ in terms of $\overline{\Psi}$. It appears that, if A and B are rather sensitive to perturbations, $\langle j_T \rangle$ is very stable and hence one can say that the profile of current density is identified in a robust way. Figure 4.5 shows a comparison between the real and identified flux lines in the plasma. Experimental data have been handled by this algorithm in [10] for the Tokamak JET, whereas different techniques have been used in [11] for the Tokamak DIII-D at General Atomics (San Diego) and in [12] for various elongated Tokamak equilibria.

We are planning to test the scaling function decomposition, as in section 3.3, for the real identification problem described in this section.

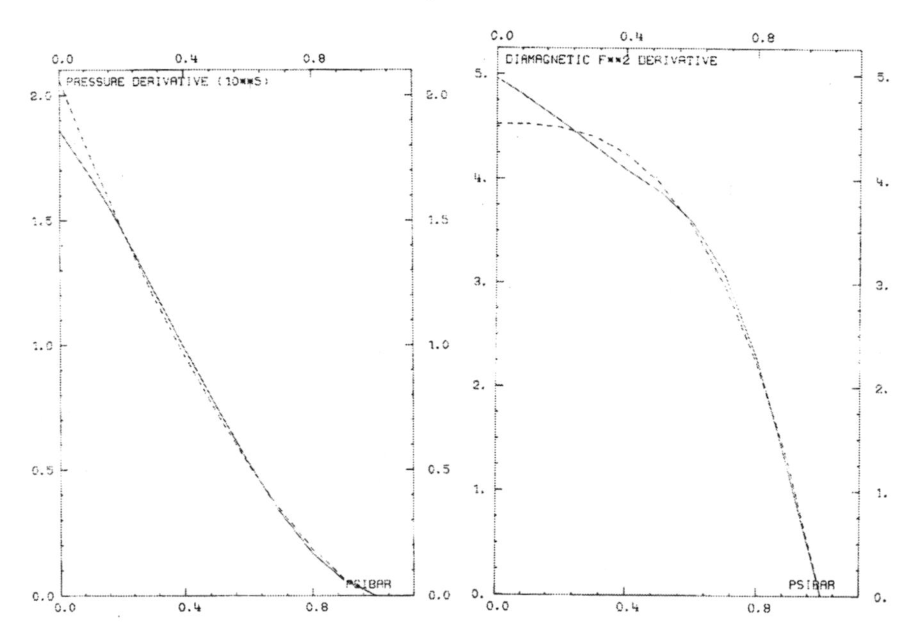

FIG. 4.3. *Identification of* $A(\overline{\Psi})$ *(left) and* $B(\overline{\Psi})$ *(right) from magnetic and polarimetric measurements*

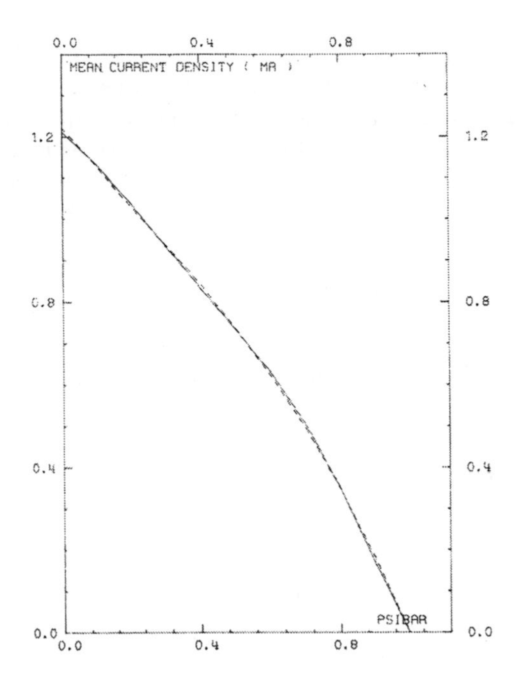

FIG. 4.4. *Profile of current density obtained by identification*

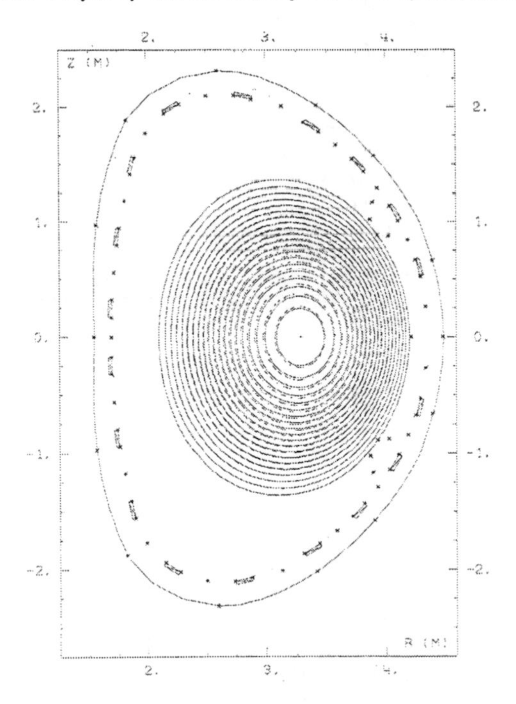

FIG. 4.5. *Comparison between real and identified flux lines in the plasma*

Acknowledgements. The authors are very thankful to M. Vogelius (Rutgers University) for many helpful discussions on the mathematical aspects of this problem and to Y. Stephan (CISI-Cadarache) for his contribution to the IDENT code. Part of these computations have been performed on the Cray C94 of CGCVR (Centre Grenoblois de Calcul Vectoriel du C.E.A.).

REFERENCES

[1] R. Gidas, W.-M. Ni, L. Nirenberg. *Symmetry and Related Properties via the Maximum Principle.* Commun. Math. Phys. 68, pp. 209–243 (1979).

[2] J. Serrin. *Asymmetry problem in potential theory.* Arch. Rat. Mech. Anal. 43, pp. 304–318 (1971).

[3] E. Beretta, M. Vogelius. *An Inverse Problem Originating from Magnetohydrodynamics.* Arch. Rat. Mech. and Anal., vol. 115, No 2, pp. 137–152 (1991).

[4] M. Vogelius. *An inverse problem for the equation* $\Delta u = -cu - d$. Annales Institut Fourier, Grenoble, 44, **4**, pp. 1181–1209 (1994).

[5] E. Beretta, E. Fischer, M. Vogelius. *An inverse problem originating from magnetohydrodynamics. Some numerical experiments.* Ill-posed Problems in Natural Sciences, ed. A. N. Tikhonov VSP, Utrecht (1992), pp. 254–269.

[6] H. Dousteyssier-Buvat. *Sur des techniques déterministes et stochastiques appliquées aux problèmes d'identification.* PhD thesis, Université Joseph Fourier, Grenoble 1 (France), 1995.

[7] I. Daubechies. *Orthonormal Bases of Compactly Supported Wavelets.* Comm. Pure Applied Math., Vol 41, 1988.

[8] S. Mallat. *Multiresolution Approximation and Wavelet Orthonormal Bases of* $L^2(\mathbb{R})$. Transaction of the A.M.S, Vol 315, $N^{\underline{o}}1$, 1989.

[9] E. Beretta, M. Vogelius. *An Inverse Problem Originating from Magnetohydrodynamics II. The case of the Grad-Shafranov equation.* Indiana Univ. Math. J., 41, pp. 1081–1118 (1992).

[10] J. Blum et al. *Problems and methods of self-consistent reconstruction of Tokamak equilibrium profiles from magnetic and polarimetric measurements.* Nuclear Fusion, Vol. 30, No 8 (1990).

[11] L.L. Lao et al. *Equilibrium analysis of current profiles in Tokamaks.* Nuclear Fusion, Vol. 30, No 6 (1990).

[12] F. Hofmann, G. Tonetti. *Tokamak equilibrium reconstruction using Faraday rotation measurements.* Nuclear Fusion, Vol. 28, No 10 (1988).

DUALITY FOR INVERSE PROBLEMS IN WAVE PROPAGATION

MARK S. GOCKENBACH* AND WILLIAM W. SYMES†

Abstract. A general dual formulation for inequality constrained optimization problems applies directly to inverse problems for multi-experiment data fitting. In the case of inverse problems in wave propagation, proper choice of the multi-experiment consistency constraint yields a dual problem with better convexity properties than the "primal" or straightforward data fitting formulation. The plane wave detection problem, a very simple inverse problem in wave propagation, provides a transparent framework in which to illustrate these ideas.

1. Introduction. Many inverse problems are naturally formulated as optimization problems, *via* minimization of a suitable data misfit measure. It is also quite common for such problems to include constraints on the model parameters. This paper is devoted to a natural constrained formulation of inverse problems arising from multi-experiments, in which a few data acquisition parameters are varied to produce multiple data sets from otherwise identical experiments. Such multi-experiments are common for example in seismology and medical imaging. It is often the case that individual experimental results can be "inverted" to yield estimates of model parameters. The natural *coherence* constraint is: all such estimates should be the same, as they are estimates of the same object (Earth or tissue, for instance).

In general, a constrained optimization problem with a single scalar constraint has a dual formulation, in which the roles of objective and constraint are exchanged. This duality arises as the result of a simple logical manipulation, and has exactly the same solution as the original problem. However since the value of the constraint at the solution is known *a priori*, in fact the two formulations have fundamentally different character.

The primal problem is to minimize the objective over the set of feasible points. The target value of the objective (that is, the noise level in the case of a data-fitting problem), is unknown *a priori*, and the solution is obtained by decreasing the objective towards its optimal value. The dual formulation involves minimizing the infeasibility (constraint violation) subject to a bound on the objective function. The target value of the constraint is known; this is part of problem formulation. As we shall show, a natural class of algorithms proceeds by *increasing* the objective (estimated noise) towards its optimal value.

In fact, we will characterize the optimal objective value (actual level of noise in the data) as the smallest zero of a certain function. In the case

* Mathematics Department, University of Michigan, Ann Arbor, MI 48109-1109.

† The Rice Inversion Project, Rice University, Department of Computational and Applied Mathematics - MS 134, 6100 Main St., Houston, Texas 77005-1892.

of a data-fitting problem, finding this zero amounts to a direct estimation of the level of noise in the data, which yields a solution of the original problem as a by-product. So we could describe the dual formulation as a "noise-finding" algorithm.

In general the dual formulation of a constrained optimization problem appears to have no intrinsic advantage from the viewpoint of computation. However we shall show that for multi-experiment inverse problems in wave propagation, the natural coherence constraint, properly formulated, can lead to dual problems which are amenable to rapidly convergent relatives of Newton's method, whereas the primal problems are not. In particular, the dual approach gives in some circumstances effective local solution methods for global nonconvex optimization problems.

The class of inverse problems which motivated this work arises in reflection seismology. We will explain in a general way the properties of the reflection seismic inverse problem which make the primal best fit problems difficult to solve with local optimization tools. Since realistic models of reflection seismology are necessarily complicated, we will instead illustrate the dual approach in detail with a very simple inverse problem in wave propagation, the plane wave detection problem. A plane wave moves through space at unit speed. It crosses a hyperplane of detectors, which sample the wavefield as a function of position on the hyperplane and time, with some error or noise. From this data one is required to reconstruct the waveform and wavevector of the plane wave. Since the measurements are presumed noisy, one will at best reconstruct estimates of these quantities. While this inverse problem is technically much simpler than those of reflection seismology, it shares the basic mathematical features which make the dual approach attractive.

In previous work on plane wave detection, ([Symes, 1994], [Gockenbach et al., 1995]), we have shown that for a particular choice of coherence constraint, regularizing the dual problem produces a problem with the properties claimed above, leading to an optimal data fit through solution of smooth convex optimization problems. Moreover we showed that the choice of coherence constraint having this favorable characteristic is essentially unique [Kim and Symes, 1996]. Here we present a more fundamental relationship between the original problem and its dual, which explains how to determine the penalty parameter over which the earlier theory gave no control, through its relation to the data noise level.

The next section describes briefly the general properties of the reflection seismic inverse problem, which it shares with other inverse problems in wave propagation in which wave velocity is to be determined. The characteristics which prevent effective application of local optimization to the primal approach will be evident. The third section defines the notion of duality used here for an arbitrary constrained optimization problem, and the fourth section specializes this notion to multi-experiment inverse problems and the coherence constraint. A further specialization to *separable* problems is useful: the plane wave detection problem, in common with many inverse problems in wave propagation, separates the model parameter into linear and nonlinear categories. This separable character leads to both theoretical and algorithmic simplifications. The final section states the plane wave detection problem and its dual formulation, explains the choice of coherence constraint leading to good behaviour of the dual problem, presents an algorithm for solution of the dual problem, and illustrates the theory with some numerical examples. The algorithm described in the last section should be quite generally applicable to dual constrained optimization problems as defined in this paper.

2. The nature of seismic reflection data. Figure 1 presents the result of a single seismic experiment in the North Sea. A ship has towed a noise-making device, in this case an array of *air guns*, which has released supersonically expanding bubbles of compressed air into the water. These in turn generated acoustic waves which propagated through the water and into the layers of rock in the subsurface. The ship also towed a cable full of special microphones, called hydrophones. Each hydrophone (or actually group) generated output voltage as the result of pressure fluctuations. Some of these pressure fluctuations were echoes, or *reflected waves*, from subsurface boundaries or other changes in rock mechanics met by the wave as it propagated. The analog—summed output voltage for each hydrophone group was digitized and recorded as a time series. These time series are the content of Figure 1.

The plotting method is typical of this subject. Time increases downwards; the time unit is milliseconds (ms). Each vertical line or *trace* represents the time varying signal recorded by a single receiver (hydrophone group). The horizontal coordinate *for* each line represents amplitude of the recorded signal (voltage, in principle). The horizontal coordinate *between* the lines represents distance from the boat or source array, in this case increasing from right to left.

The variations in amplitude are evidently oscillatory. The plot has been filled with black ink in each excursion of the amplitude to the right of the axis in each trace. This device brings out another characteristic of the data: the presence of *space time coherent signals*, known as *reflections*. Evidently certain signal components occur at certain times on each trace, and as one moves away from the boat the time of occurrence becomes later in a systematic way. It is natural to think that this time represents distance traveled by the wave, and this idea can be justified through appropriate mathematical models of wave propagation and reflection.

Not all of the data is displayed in Figure 1. Sample values have been replaced by zeroes above a line sloping down to the left. The reason for this cutoff or *mute* is that part of the data above this line represents a different physical regime and has a different scale.

This single seismic experiment produced 120 traces, each with 750 samples, representing 3 s of recorded data. (For other dimensional information see the figure caption.) However the data set from which Figure 1 was extracted contained more than 1000 such experiments. The ship steamed along a line, and conducted one experiment (a "shot") every ten seconds or so. The data collectively represents the influence of the subsurface of the earth to roughly 3 km, over a narrow swath roughly 25 km long through the North Sea. The total volume of data is roughly 750 Mb.

Moreover this is a small data set. The geometry of acquisition (ship sailing in a single line towing one cable) is termed "2D" in the current seismic literature. Almost all contemporary acquisition is in "3D" mode, which in practice means (at least) a ship towing up to a dozen cables steaming successively along many parallel lines. Data volumes of hundreds of Gbytes or even Tbytes are not unusual. Evidently data processing methods of considerable efficiency are required to extract from this vast quantity of data useful information about the subsurface.

Because the data of reflection seismic experiments contain reflections (space-time coherent oscillatory signals), the *phase* of these signals is of particular importance - that is, the relation between their time of arrival and the source-receiver distance. The simplest models of seismic wave propagation having this important feature - generation of reflections with definite and realistic phase - is the so called primaries only or linearized acoustic model, and its variants. This model parametrizes Earth mechanics by a smooth or slowly varying *velocity* field, heterogeneous on a scale of $10^2 - 10^3$ m, and a rapidly varying *reflectivity* field, heterogeneous on a scale of $10^0 - 10^1$ m. Thus the scales of parameter variation are *separated into two regimes, which are treated as different fields.* Roughly, the reflectivity r is responsible for the amplitudes of reflected waves, and the velocity v for the phase or arrival time.

To obtain this model, one begins with the acoustic wave equation, thus neglecting a variety of important physics (eg. that the Earth is not a fluid!). Several stages of approximation lead to the representation of the seismogram as an oscillatory integral:

$$S[v,r](x_s,x_r,t) =$$

$$\int_{\text{Earth}} dx \int_{\mathbf{R}^3} d\xi a[v](x_s,x_r,t,x,\xi)\, r(x) \exp\left(i[t - \phi[v](x_s,x_r,x,\xi) + x\cdot\xi]\right)$$

Here

- x_s = source location
- x_r = receiver location
- t = time
- x = location in Earth model (space)
- ξ = spatial frequency (vector)
- $v(x)$ = velocity field

- $r(x)$ = reflectivity field
- $a[v]$ = amplitude
- $\phi[v]$ = phase

The phase $\phi[v]$ is positively homogeneous of degree 1 in the frequency variable ξ. The amplitude $a[v]$ has an asymptotic expansion in decreasing powers of $|\xi|$ at infinity. These functions depend on the velocity v through the constructions of *geometric optics*. For accounts of the steps leading to this expression and further references see [Beylkin, 1985], [Virieux et al., 1992], [Sevink and Herman, 1994], [Symes, 1991], and references cited in these papers.

Trouble comes through the dependence of the phase ϕ on v. Small changes in v give changes in ϕ proportional to $|\xi|$ (since ϕ is homogeneous of degree one in ξ), which are large at high spatial frequencies $|\xi|$. This effect is quite visible in simulations. Figure 2 shows a simulated seismogram using a moderately realistic model (choice of v, r, and other parameters). The simulation was carried out through direct numerical evaluation of the oscillatory integral formula described above. Figure 3 shows another simulated seismogram, with the same parameters as that of Figure 2 except that the velocity v has been decreased by 2%. Figure 4 shows the difference between the data of Figures 2 and 3, plotted on the same scale: it is bigger than either summand. Note that the overall size of the seismogram has not changed between Figures 2 and 3, only the phases of the oscillatory components. Thus the very rapid change in the seismogram $S[v,r]$ represents very nonlinear behaviour.

A natural best fit formulation of the seismic inverse problem is: adjust v, r to fit the data as well as possible in the mean square sense, i.e.

$$\min_{v,r} \ ||S[v,r] - S_{\text{data}}||^2$$

where $||\cdot||$ is an L^2 type norm (integration over x_s, x_r, and t, perhaps with a weight) and S_{data} is measured data to be fit.

Because of the extremely nonlinear dependence of S on v, this functional tends to behave very nonquadratically. In fact many researchers have assumed it to be highly multimodal, with many spurious local minima (see eg. [Scales et al., 1991]). While existence of multiple local minima has actually been established in only a few instances, experience clearly indicates that this is indeed an extremely difficult objective to minimize using descent methods. In view of the analysis outlined above, this is easy to understand: quadratic models would necessarily have very small domains of accuracy. For some computational experience, see [Gauthier et al., 1986], [Kolb et al., 1986], [Mora, 1986], and [Santosa and Symes, 1989].

Evidently the reflection seismic inverse problem has a rather complex structure, including many features which we have not emphasized in this very brief account. This complexity makes the analysis of the least squares objective function defined above rather difficult. Several years ago, the second author introduced a very simple model problem, the *plane wave detection problem*, as possibly the simplest problem having the same general features as the reflection seismic velocity estimation problem. We will use this model problem to illustrate the duality approach we propose to inverse problems in wave propagation, to the definition of which we now turn.

3. A duality principle for constrained optimization. The duality explored in this paper makes sense for an arbitrary optimization problem with a scalar constraint:

$$\min \{f(x) : h(x) \leq 0\} \tag{3.1}$$

Here $f,\, h : X \to \mathbf{R}$ are continuously differentiable scalar functions on the Hilbert space X. We assume that

$$\{x \in X : h(x) \leq 0\}$$

is nonempty, and that

$$\inf \{f(x) : x \in X, h(x) \leq 0\} > \inf \{f(x) : x \in X\}.$$

The dual problem arises from an alternate expression for the constrained minimum:

$$\begin{aligned} \min_{x \in X} \{f(x) : h(x) \leq 0\} &= \inf \ \{\epsilon : \{x : f(x) \leq \epsilon\} \cap \{x : h(x) \leq 0\} \neq \emptyset\} \\ &= \sup \{\epsilon : \{x : f(x) \leq \epsilon\} \cap \{x : h(x) \leq 0\} = \emptyset\} \\ &= \sup \{\epsilon : \inf \ \{f(x) : h(x) \leq 0\} > \epsilon\} \\ &= \sup \{\epsilon : \inf \ \{h(x) : f(x) \leq \epsilon\} > 0\} \end{aligned}$$

We wish to apply this equivalence to constrained least-squares problems of the form

$$\begin{aligned} \min \quad & \tfrac{1}{2}\|F(x) - d\|^2 \\ s.t. \quad & G(x) = 0 \end{aligned} \;, \tag{3.2}$$

where $G : X \to Y$ and Y is another Hilbert space. Define

$$f(x) = \frac{1}{2}\|F(x) - d\|^2, \quad h(x) = \frac{1}{2}\|G(x)\|^2$$

and $H : \mathbf{R} \to (-\infty, +\infty]$ by

$$H(\epsilon) = \inf \ \{h(x) : f(x) \leq \epsilon\} \tag{3.3}$$

Then there exists an interval (ϵ_0, ϵ^*) such that
* $H(\epsilon) = +\infty \ \forall \ \epsilon < \epsilon_0 \ (\epsilon_0 \geq 0)$;
* $H(\epsilon) > 0 \ \forall \ \epsilon \in (\epsilon_0, \epsilon^*)$;

* $H(\epsilon) = 0 \ \forall \ \epsilon \geq \epsilon^*$.
* H is monotone decreasing.

Moreover, the desired constrained minimum is ϵ^*, that is, to find the optimal value of (3.2), we must locate the first zero of H.

Suppose that a continuous path of solutions $x(\epsilon)$ of the optimization problems defining H in (3.3) exists, along with a corresponding path $\lambda(\epsilon)$ of Lagrange multipliers. The following conditions hold:

$$\begin{aligned} \nabla h(x(\epsilon)) + \lambda(\epsilon)\nabla f(x(\epsilon)) &= 0; \\ \lambda(\epsilon) &\geq 0; \\ \lambda(\epsilon)\left(f(x(\epsilon)) - \epsilon\right) &= 0; \\ H(\epsilon) &= h(x(\epsilon)). \end{aligned}$$

From these conditions, it is easy to derive

$$H'(\epsilon) = -\lambda(\epsilon).$$

Now, H is related to the quadratic penalty function for problem (3.2):

$$P(x;\rho) = \frac{1}{2}||F(x) - d||^2 + \frac{\rho}{2}||G(x)||^2.$$

In particular, if $y(\rho)$ is a minimizer of $P(\cdot;\rho)$, then $x(\epsilon) = y(\rho)$ is a solution of (3.3), where

$$\epsilon = f(y(\rho));$$

moreover,

$$\lambda(\epsilon) = \rho^{-1}.$$

It follows from the standard theory of penalty functions [Fiacco and McCormick, 1990] that $H'(\epsilon) = -\rho^{-1}$ is increasing as a function of ϵ, and so
* H is convex.

From this we also can deduce that
* $H'(\epsilon^*) = 0$.

We now have a rather complete description of H. The last property is a drawback; we would like to use an efficient zero-finding technique to find ϵ^*, but fast convergence of Newton's method and similar algorithms is precluded by the fact that the root is also a stationary point. We can easily overcome this problem, though, by defining

$$\tilde{H}(\epsilon) = \sqrt{2H(\epsilon)};$$

then

$$\tilde{H}'(\epsilon) = \frac{H'(\epsilon)}{\sqrt{2H(\epsilon)}} = -\frac{1}{\rho||G(y(\rho))||}.$$

By the standard theory of the quadratic penalty function,

$$\rho||G(y(\rho))|| \to ||\bar{\lambda}||,$$

where $\bar{\lambda}$ is the Lagrange multiplier for problem (3.2). Thus we obtain, in the typical case in which $\bar{\lambda} \neq 0$,
* $\tilde{H}(\epsilon^*) = 0, \; \tilde{H}'(\epsilon^*) < 0.$
This allows the efficient location of ϵ^*.

A simple illustration of this concept is provided by linear regression in the plane. Given vectors $a, \xi \in \mathbf{R}^2$, we are to find the closest vector to a on the line determined by $\xi^T x = 0$. The "primal" problem is

$$\min_x \left\{ |a - x|^2 : |\xi^T x|^2 \leq 0 \right\}$$

and the dual formulation seeks the smallest zero of

$$H(\epsilon) \equiv \inf_x \; \left\{ |\xi^T x|^2 : |a - x| \leq \epsilon \right\}$$

A short calculation shows that

$$H(\epsilon) = \begin{cases} (|\xi^T a| - \epsilon)^2, & \epsilon < |\xi^T a| \\ 0 & \text{otherwise} \end{cases},$$

whence $\epsilon^* = |\xi^T a|$. Note that H has the properties discussed above.

Note that for values of ϵ near ϵ_0, the computation of H involves the solution of a problem that approximates the following problem:

$$\begin{aligned} & \min \quad \tfrac{1}{2}||G(x)||^2 \\ & s.t. \quad \tfrac{1}{2}||F(x) - d||^2 \leq \epsilon_0 \end{aligned} \tag{3.4}$$

It is problem (3.4) that we call the dual problem; for small values of $\epsilon - \epsilon_0$, computing $H(\epsilon)$ involves solving a regularized version of (3.4). Note that in the case that the data can be fit exactly in the absence of the condition $G(x) = 0$ (that is, in the case that $\epsilon_0 = 0$), the constraint in (3.4) becomes $F(x) = d$; in either case, the role of the objective function and constraint have been exchanged.

Since computing $H(\epsilon)$ also involves solving a problem with the objective and constraint exchanged, we shall refer to the problem of finding the smallest zero of H as the dual formulation or dual approach.

In general the dual formulation appears to be a step backwards: instead of solving a constrained optimization problem, we must solve an equation involving a function defined by solution of a family of similar problems depending on a parameter. In order for this transformation to yield an algorithmic advantage, the problems defining H must be in some sense easier to solve than the original.

We shall show later that in the case of a multi-experiment inverse problem, the dual problem consists of minimizing the incoherence in the model estimates obtained by inverting the data from each experiment separately. For a certain class of problems arising in wave propagation, when the coherence constraint is chosen properly, this dual problem has global properties which allow it to be solved using efficient local optimization techniques.

4. Multi-experiment inverse problems. An inverse problem seeks to determine model parameters $m \in M$ from data measurements $d \in D$ through an assumed relationship $d \simeq F(m)$. We will assume that the model and data spaces M and D are open sets in Hilbert spaces, and use $||\cdot||$ to denote the norm in either. Then a natural "best fit" formulation of such an inverse problem is to seek $m \in M$ to minimize the mean square error $||F(m) - d||^2$.

In many such problems the model parameter vector m is subject to various constraints, either of physical origin or imposed to render the misfit minimization better behaved computationally. Here we introduce another class of constraints, which appear naturally when the measurements are the result of *multi-experiments.* As explained in the introduction, a multi-experiment is a parameterized set of similar experiments, explicable by a common model and producing data of the same mathematical type. Denote by e the experimental parameters. Then a simple formalism encompassing both discrete and (idealized) continuous multi-experiments posits that $e \in E$, where $(E, d\mu)$ is a measure space. For discretely parameterized experiments (always the case in real world measurement) $d\mu$ is a discrete measure of finite support.

A natural *multi-experiment data space* is then $D \equiv L^2(E, K, d\mu)$, where K is the data space for a single experiment.

Since each experiment is independent of the others, we can also introduce a *multi-experiment model space*

$$X \equiv L^2(E, M, d\mu)$$

(or a subspace carrying a stronger topology—see the example given below), and by abuse of notation also write $F : X \to D$. That is, we will write $F(x)(e)$ for $F(x(e))$. This expresses the hypothesis that each experiment must be explicable by the physics embodied in the model-data relation F.

The basic model space M is isomorphic to the *coherent* members of X, i.e. those $x \in X$ for which $x(e_1) = x(e_2)$ for μ-almost every $e_1, e_2 \in E$. The coherent models form a closed subspace of X isomorphic to M, and we will write $M \subset X$. A *coherence operator* W is a continuous function $W : X \to Z$, Z another Hilbert space, for which $M \subset \{x \in X : W(x) = 0\}$. This is a rather relaxed notion: for example $W = 0$ satisfies it, and is of course a useless choice. If the stronger statement $M = \{x \in X : W(x) = 0\}$ holds, then the original best fit formulation of the inverse problem

$$\min\left\{||F(m) - d||^2 : m \in M\right\}$$

is equivalent to the constrained problem

$$\min\left\{||F(x) - d||^2 : x \in X, W(x) = 0\right\}.$$

However it will turn out to be sufficient to study this constrained problem for operators W satisfying only the weaker notion of coherency. If the data are noise free, that is $d = F(x^*)$ for some $x^* \in M$, then x^* is a solution of both problems. *Thus we shall simply replace the original best fit formulation of the inverse problem by the constrained optimization problem just stated.*

5. Separable multi-experiment problems. Separable means here that the *model parameters* divide into two classes, those having a linear influence on the data and those whose influence is nonlinear. By an abuse of language we shall refer to these as the linear and nonlinear parameters, respectively. The nonlinear parameters vary over a set S. In order that calculus operations be well-defined, we assume that S is an open set in a separable Hilbert space. The linear parameters range over another separable Hilbert space Y.

The measure space $\{E, d\mu\}$ parameterizes the multiple experiments: there is one experiment for each $e \in E$. As before, the *multiexperiment data space* is

$$D \equiv L^2(E, K, d\mu)$$

In the applications with which we are familiar, there is no need to make the nonlinear parameters experiment-dependent; the reason for this is that the data can be well-explained as long as the linear parameter varies with the experiment. Thus, only the linear model parameters will be made experiment-dependent. The *multiexperiment model space* is

$$X = S \times U,\ U \equiv L^2(E, Y, d\mu)$$

or a subspace. As before the coherent models, i.e. those $x = (s, u) \in X$ for which $u(e_1) = u(e_2)$ for μ-almost all $e_1, e_2 \in E$, form a subspace of X which we identify with $M \equiv S \times Y$.

We will consider only coherence operators $W : X \to Z$ which are linear in the second argument, and write $W(s)u$ for the image.

The problem is separable exactly because the forward map $F : X = S \times U \to D$ is linear in the second argument, which we reflect in the notation:

$$F(s, u) = A(s)u,\ A(s) \in L^\infty(E, \mathcal{L}(Y, K), d\mu) \subset \mathcal{L}(U, D)$$

where $\mathcal{L}$ denotes the space of bounded linear maps with domain and range as indicated.

With this notation, the primal statement of the inverse problem is:

$$\min\left\{||A(s)u-d||^2 : (s,u)\in X, W(s)u=0\right\}. \tag{5.1}$$

The dual problem is

$$\min\left\{\frac{1}{2}||W(s)u||^2 : A(s)^*A(s)u = A(s)^*d\right\}.$$

To solve the primal problem, we find the first zero of

$$H(\epsilon)\equiv\inf\left\{\frac{1}{2}||W(s)u||^2 : \frac{1}{2}||A(s)u-d||^2\leq\frac{1}{2}\epsilon^2\right\}.$$

6. Plane wave detection.

6.1. Primal formulation. A plane wave (in the plane, i.e. 2 + 1 dimensional space-time with coordinates (t,x,y)) is a function of the form $u(t-x\sin\theta-y\cos\theta)$.

The plane wave detection problem is: given samples of a plane wave along a line in the plane, determine the incidence angle and waveform. This problem is a caricature of an important inverse problem in ocean acoustics. However it is mostly interesting as a simple model of more complex inverse problems in wave propagation, with which it shares fundamental mathematical similarities.

The line along which samples are measured might as well be the x-axis; then in the ideal case of continuous measurements over intervals in t and x, the data are

$$\{u(t-sx) : (t,x)\in\Omega=[t_{\min},t_{\max}]\times[x_{\min},x_{\max}]\},$$

where $s=\sin\theta$. If u represents acoustic pressure, for instance, then on physical grounds u should be square-integrable (as its square is proportional to the power flux density across the x-axis in that case).

The unknowns of the plane wave detection problem are the direction sine s and the waveform $u(t)$. This is a separable multi-experiment inverse problem in the sense of the previous section, as we shall now see.

The experiment space E is the interval $[x_{\min},x_{\max}]$, as we can regard each sensor as recording a possibly independent experiment. To conform to the notation already introduced, we will write $u(t-se)$ for the predicted signal at sensor location e. Take for the measure μ on E Lebesgue measure, in some ways the simplest choice.

The single experiment data space is $K=L^2([t_{\min},t_{\max}])$. In order to deal with technical difficulties, the model-data map will be defined by

$$A(s)u(t,e) = u'(t - se),$$

with the single experiment model space M equal to $S \times Y$, where

$$Y = \left\{ u \in H^1[0,T] : \int_{t_{\min}}^{t_{\max}} u = 0, u(T) = u(0) \right\}.$$

(So note that, having selected some $\tilde{u}$, $\tilde{s}$ as best explaining the data, the prediction of the waveform will be the time-derivative of $\tilde{u}$). We interpret the shift in the forward operator periodically. The multi-experiment model space will be taken to be

$$X \equiv S \times U, \; U = H^1(E, K, de) \cap L^2(E, Y, de)$$

and the multi-experiment version of the forward map is

$$A(s)u(t,e) = \frac{\partial u}{\partial t}(t - se, e).$$

Evidently the space of nonlinear variables s is $S = (-1, 1)$, since $s = \sin\theta$. Define the *output least-squares* objective function

$$J_{OLS} : (-1,1) \to \mathbf{R}$$

by

$$J_{OLS}(s) = \min \left\{ \frac{1}{2} \|A(s)u - z\|^2 : u \in M \right\},$$

where $z \in X$ is the observed data.

Examination of J_{OLS} for coherent models, i.e. the straightforward least squares formulation, reveals the most important single fact about the plane wave detection problem: its objective function is not smooth. Indeed, the mapping

$$(s, u) \in \mathbf{R} \times Y \mapsto u'(\cdot - se) \in L^2([t_{\min}, t_{\max}])$$

is not even locally uniformly continuous (for $e \neq 0$). The plane wave forward map defined here shares this property with more complex data prediction operators for wave propagation.

Another, related property of the translation map, as defined here, is the instability of its range: for $s_1 \neq s_2$, the ranges of $A(s_1)$ and $A(s_2)$, restricted to coherent models, are essentially orthogonal. Indeed, these ranges consist of plane waves with different slopes. Consequently the objective function can change arbitrarily rapidly; it tends to rise quickly from its global minimum, and assume a plateau value.

To illustrate this, we performed the following synthetic experiment. We chose for u^* the wavelet whose derivative is displayed in Figure 5, and chose $s^* = 0.2$. With 41 receivers evenly distributed on the interval $[-0.5, 0.5]$, we simulated 2 seconds of data, and then added filtered Gaussian noise (all computations here and below were done using discrete Fourier transforms in time and finite differences in space). The result is the data shown in Figure 6.

We then computed the function J_{OLS} on the interval $(-1, 1)$; this function is plotted in Figure 7. Clearly no local optimization technique will be able to find the desired minimum unless a very good estimate of the solution is available to use as a starting point.

6.2. Dual formulation. It is possible to show that essentially the only coherence operator which yields a smooth dual objective function for the plane wave detection problem is the partial derivative with respect to the experiment parameter [Kim and Symes, 1996]. We therefore choose

$$W = \frac{\partial}{\partial e}.$$

It is easy to show that if there is no noise in the data, then the objective function for the dual problem is

$$J_0(s) = \text{const}(s - s^*)^2.$$

Full details concerning this computation may be found in [Gockenbach et al., 1995].

Since the problem defining $H(\epsilon)$ approximates the dual problem when ϵ is small, we can expect that it will be easy to compute H for small ϵ. We define

$$J_\epsilon(s) = \min_u \left\{ \frac{1}{2} \|Wu\|^2 : \|A(s)u - z\| \leq \epsilon \right\},$$

so that

$$H(\epsilon) = \min \left\{ J_\epsilon(s) : s \in (-1, 1) \right\}.$$

As we pointed out in Section 3, there is a close relationship between the problem of minimizing J_ϵ and the problem of minimizing $\tilde{J}_\sigma$ defined by

$$\tilde{J}_\sigma(s) = \min \left\{ \frac{1}{2} \|A(s)u - z\|^2 + \frac{\sigma^2}{2} \|Wu\|^2 : u \in U \right\}.$$

The function J_σ has been extensively studied; the results quoted below may be found in [Gockenbach et al., 1995].

We first note the obvious result that if the data is exact, that is, if $z = A(s^*)u^*$ for some $s^* \in (-1, 1)$ and $u^* \in M$, then s^* minimizes $\tilde{J}_\sigma$

for all positive σ. We wish to know that this solution is perturbed in a controlled fashion when the data is perturbed.

Because J_{OLS} is not smooth for general $z \in K$, we must restrict the data to have some smoothness in order to allow the use of the implicit function theorem and other calculus-based theory. This in turn allows strong conclusions to be drawn about the effect of noise in the data on J_σ. To this end, we introduce the Sobolev-based spaces

$$Y_k = \left\{ u \in H^k[0,T] : \int_0^T u = 0, \frac{\partial^i u}{\partial t^i}(T) = \frac{\partial^i u}{\partial t^i}(0), i = 0, \ldots, k-1 \right\},$$

$$K_k = Y_k,$$

and

$$U_k = \cap_{j=0}^k H^j(E, K_{k-j}, de).$$

We then have the following results.

THEOREM 6.1. *If $z^* = A(s^*)u^*$, where $u^* \in Y_5$ and $s^* \in (-1,1)$, then for any $\eta > 0$, there exist $\sigma_0 > 0$, and $\delta > 0$ such that if $z \in U_4$ and $\|z - z^*\|_{U_2} < \delta$, then there exists a differentiable path $\tilde{s}_\sigma$, parameterized by $\sigma \in (\sigma_0, +\infty)$, such that $\tilde{s}_\sigma$ is a local minimizer of $\tilde{J}_\sigma$ and $|\tilde{s}_\sigma - s^*| < \eta$.*

COROLLARY 6.1. *Under the hypotheses of the preceding theorem, there exists a continuous path s_ϵ, defined on (ϵ_1, ϵ^*), such that s_ϵ is a minimizer of J_ϵ and $s_\epsilon = \tilde{s}_\sigma$ when $\epsilon = \|A(\tilde{s}_\sigma) - z\|$.*

We can therefore apply the general theory discussed in Section 2, with two caveats. First of all, we do not know that a Lagrange multiplier exists for the output least-squares problem

$$\begin{array}{ll} \min & \frac{1}{2}\|A(s)u - z\|^2 \\ s.t. & Wu = 0 \end{array} . \tag{6.1}$$

This is an infinite dimensional constrained optimization problem and the linear constraint operator does not satisfy the standard regularity condition that guarantees the existence of a Lagrange multiplier. In the case of noise-free data, the multiplier does exist and can be computed explicitly, but when the data contains noise, its existence is an open question. The existence of a nonzero multiplier was used in Section 2 to guarantee that the function $\tilde{H}$ satisfied $\tilde{H}'(\epsilon^*) \neq 0$. Secondly, there is no guarantee that $\epsilon_1 = \epsilon_0$; in other words, there is no guarantee in the above theorem that the path s_σ extends as σ gets arbitrarily close to zero.

Remark. We have recently discovered that a different definition of the operator W removes the second obstacle mentioned in the preceding paragraph [Kim and Symes, 1996]. We have not yet found a way around the first.

We now demonstrate the nature of J_ϵ. In Figure 8, we display the graphs of this function for four values of ϵ. As expected, for small values of ϵ, J_ϵ resembles J_0, which is quadratic.

These plots should be compared to Figure 7; obviously it is simple to minimize J_ϵ for a small value of ϵ.

The function $\tilde{H}$ is plotted in Figure 9.

This function is almost linear; it is easy to find the desired root using Newton's method; recall that the derivative of $\tilde{H}$ is available, since it is related to the Lagrange multiplier for the optimization problem defining $\tilde{H}$. As indicated in Figure 9, only a few iterations are required to find ϵ^* to good accuracy.

It should be noted that although, for this example, J_ϵ becomes more nonconvex as ϵ increases, we can expect to have a very good starting point for minimizing J_ϵ after we have done it once; moreover, the first minimization, for a small value of ϵ, is simple. Therefore it possible both to find an initial point on the path s_ϵ and to follow it as $\epsilon \to \epsilon^*$.

6.3. Computing J_ϵ. We close by returning to the general separable multi-experiment inverse problem (5.1). In analogy to what we did for the plane wave detection problem, we can define

$$J_\epsilon(s) = \min\left\{\frac{1}{2}||W(s)u||^2 : ||A(s)u - d|| \leq \epsilon\right\},$$

so that

$$H(\epsilon) = \min\left\{J_\epsilon(s) : s \in S\right\}.$$

The problem of minimizing J_ϵ is likely to require different algorithms for different applications (for instance, the description of the set S is likely to influence the choice of algorithm). However, we can suggest an algorithm for *computing* $J_\epsilon(s)$ for a given s, that is, for solving the quadratic-quadratic problem

$$\begin{array}{ll} \min & \frac{1}{2}||Wu||^2 \\ s.t & \frac{1}{2}||Au - d||^2 \leq \frac{1}{2}\epsilon^2 \end{array} \quad (6.2)$$

Problem (6.2) is similar to the trust-region subproblem that arises in unconstrained optimization (see [Dennis and Schnabel, 1983]); we propose to adapt the algorithm commonly known as the Hebden-Moré algorithm (see [Hebden, 1973], [Moré, 1977]). Since (6.2) is a convex problem, it is equivalent to its first-order optimality conditions: find $\mu \in \mathbf{R}$, u such that

$$(\mu^2 A^* A + W^* W)\, u = \mu^2 A^* d. \tag{6.3}$$

$$\mu \left(\frac{1}{2} ||Au - d||^2 - \frac{1}{2} \epsilon^2 \right) = 0 \tag{6.4}$$

$$\mu \geq 0 \tag{6.5}$$

We define u_μ to be the solution of (6.3) for a given μ and $r(\mu) = ||Au_\mu - d||$.

Algorithm for solving (6.2):

Given upper and lower bounds μ_{up} and μ_{low} for μ, a tolerance $\eta_1 > 0$ for $|\frac{1}{2}||Au - d||^2 - \frac{1}{2}\epsilon^2|$, and a tolerance $\eta_2 > 0$ for $||Wu||$:

1. Choose $\mu \in [\mu_{low}, \mu_{up}]$.
2. do
    1. Compute u_μ and $r(\mu)$;
    2. Compute $J = ||Wu_\mu||$;
    3. if $(r(\mu) < \epsilon$ and $J/||u_\mu|| < \eta_2)$ or $(|r(\mu) - \epsilon| < \eta_1 \epsilon)$
        - DONE
    4. else
        1. Compute $r'(\mu)$ and choose a and b so that

            $$\phi(\mu) = r(\mu) \text{ and } \phi'(\mu) = r'(\mu),$$

            where

            $$\phi(\eta) = \frac{a}{\eta^2 + b}.$$

        2. Find μ_+ by solving $\phi(\eta) = 0$;
        3. if $(\mu = \mu_{low}$ and $\mu_+ < \mu_{low})$ or $(\mu = \mu_{up}$ and $\mu_+ > \mu_{up})$
            - DONE
        4. else if $\mu_+ < \mu_{low}$
            - set $\mu = \mu_{low}$
        5. else if $\mu_+ > \mu_{up}$
            - set $\mu = \mu_{up}$
        6. else set $\mu = \mu_+$.
    5. while NOT DONE

Note that this algorithm is based on approximating the function r by a function of the form

$$\phi(\eta) = \frac{a}{\eta^2 + b}.$$

It costs two linear solves per iteration (to compute u_μ in step 2(a) and to compute $r'(\mu)$ in step 2(d)(i)). Since the operator in both linear equations is the same, when direct solves are employed, this requires one factorization per iteration; when iterative solves are used, the efficiency depends on having a good preconditioner.

For this algorithm, we require that, for $\mu \neq 0$, the operator

$$\mu^2 A^* A + W^* W \tag{6.6}$$

is positive definite. If this is not the case, it implies that all of the problems under consideration are ill-posed. In this case, a regularization operator R may be chosen, and the primal problem posed as

$$\min \left\{ ||A(s)u - d||^2 : (s,u) \in X, W(s)u = 0\,, ||Ru|| \leq \delta \right\}. \tag{6.7}$$

In this case, the above algorithm could be modified to adjust two Lagrange multipliers to satisfy the corresponding first-order conditions.

We close by noting that in the case of the plane wave detection problem, we were able to pose the problem in such a way that operator (6.6) was a scaled Laplacian; we were thus able to use a Fast Poisson method in the above algorithm.

Acknowledgement and Disclaimer. This work was partially supported by the National Science Foundation under grant number DMS 9404283; the Office of Naval Research under grant number N00014-96-1-0156; the Air Force Office of Scientific Research, Air Force Materials Command, USAF, under grant number F49620-95-1-0339; the Schlumberger Foundation; and The Rice Inversion Project. TRIP Sponsors for 1996 are Advance Geophysical, Amoco Production Co., Conoco Inc., Cray Research Inc., Discovery Bay, Exxon Production Research Co., Interactive Network Technologies, Mobil Research and Development Corp., and Shell International Research. The US Government is authorized to reproduce and distribute reprints for governmental purposes notwithstanding any copyright notation thereon. The views and conclusions contained herein are those of the authors and should not be interpreted as necessarily representing the official policies or endorsements, either expressed or implied, of the Air Force Office of Scientific Research, the Office of Naval Research, the National Science Foundation, or the US Government.

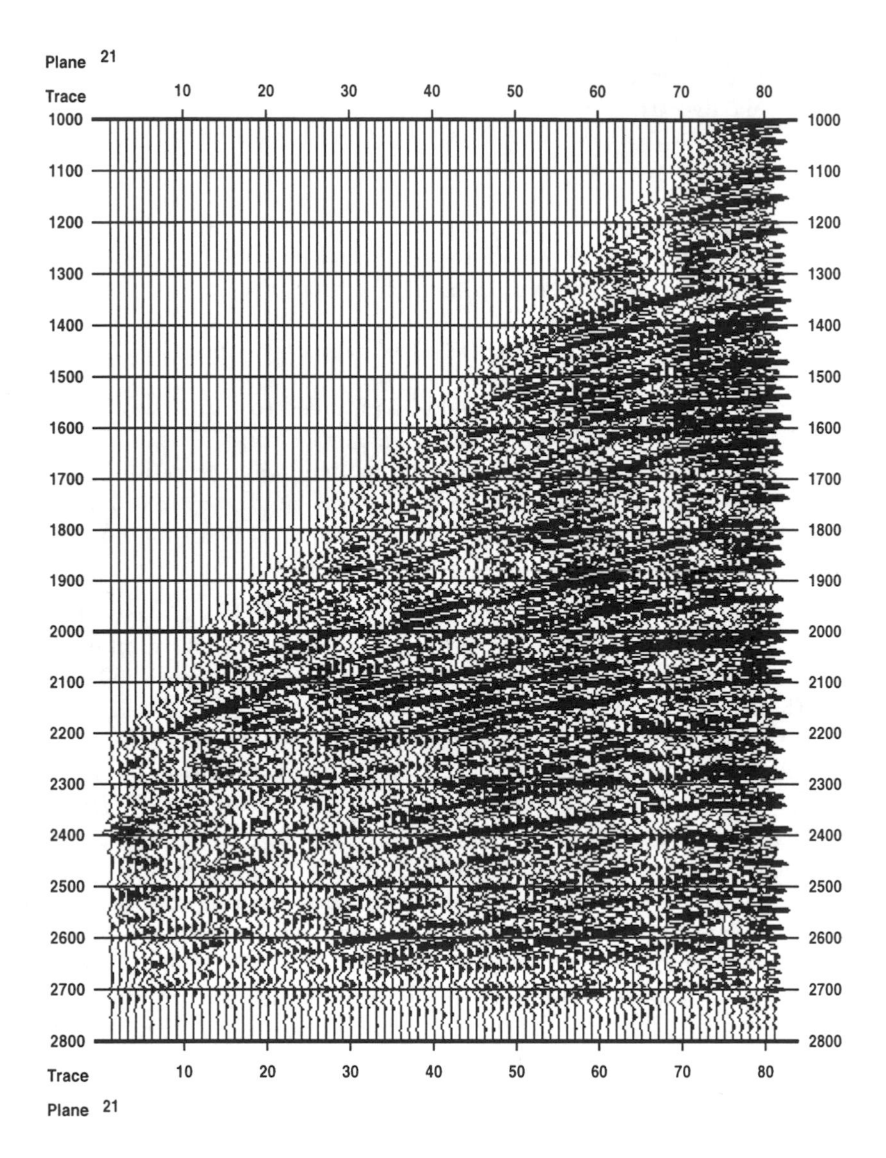

FIG. 1. *Shot record (single experiment data) from seismic data set acquired in the North Sea. Time sample rate is 4 ms, total time of recording was 3 s. Hydrophone group spacing is 25 m. Closest offset (distance from source to receiver) is 262 m, far offset is 2262 m. Tow depth of source array was 10 m, depth of hydrophone cable was 6 m. The authors gratefully acknowledge Mobil Research and Development Corp. for provision of this data and permission for its use.*

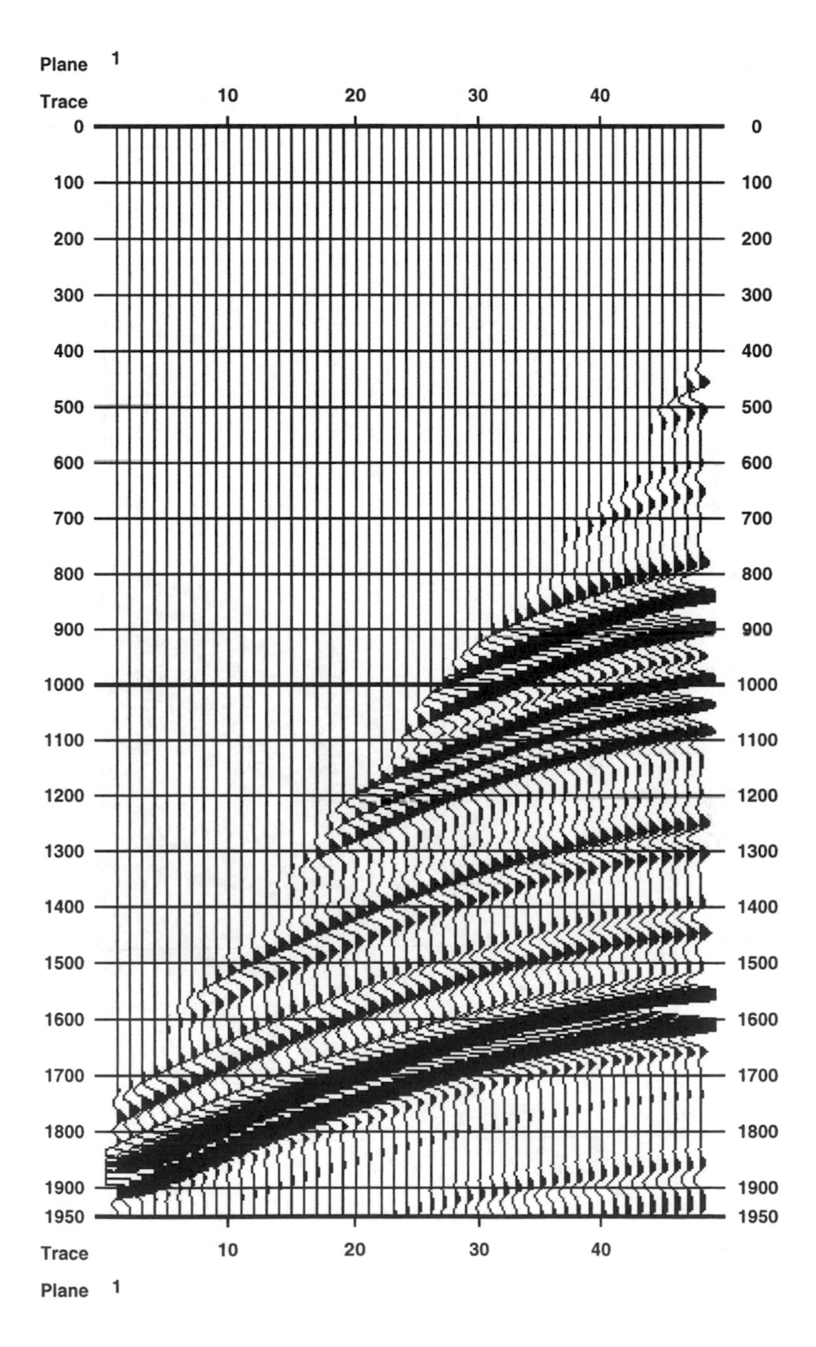

FIG. 2. *Simulated muted shot record. Method of simulation was finite difference approximation of the system of wave equations defining the linearized acoustic model.*

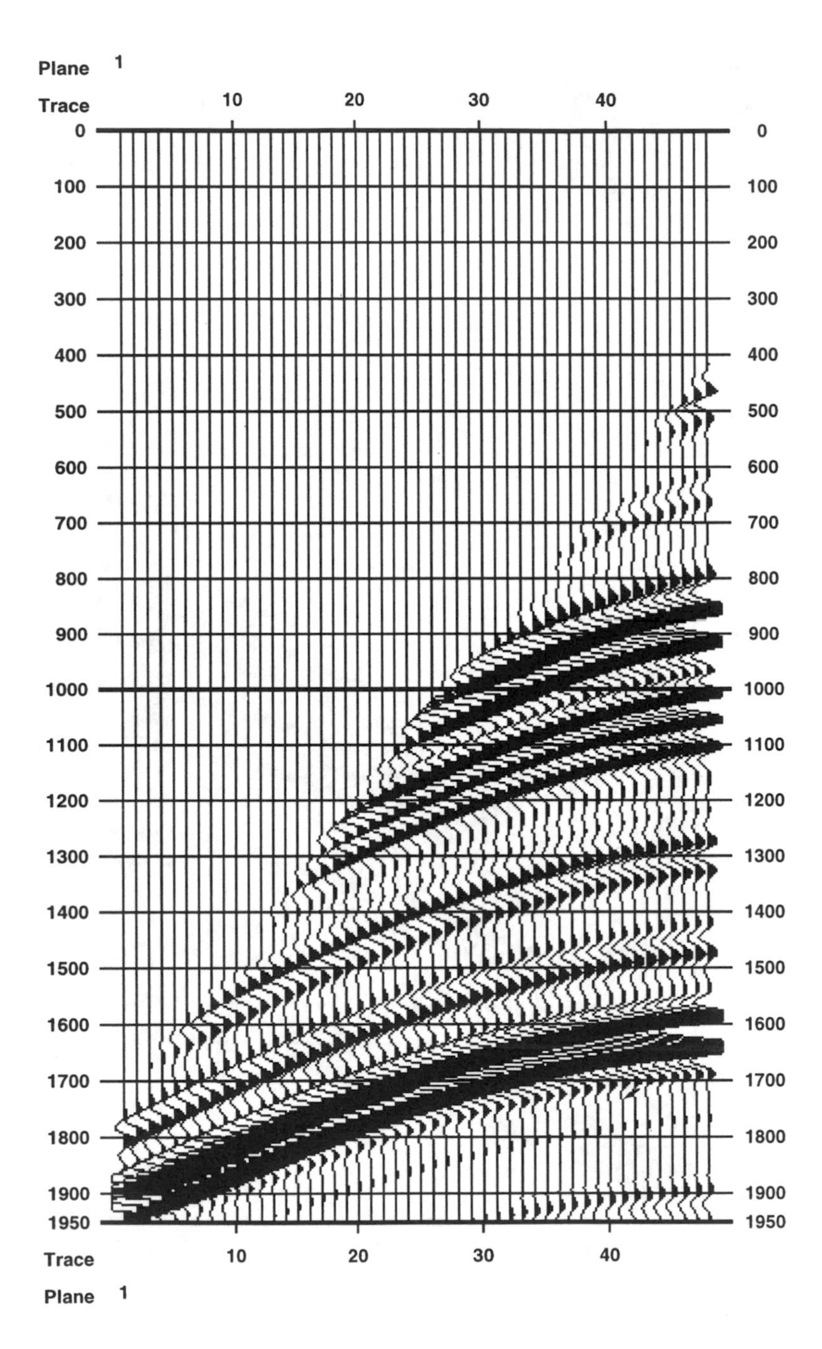

FIG. 3. *Simulated muted shot record. Velocity field uniformly 2% lower than that used to produce Figure 2.*

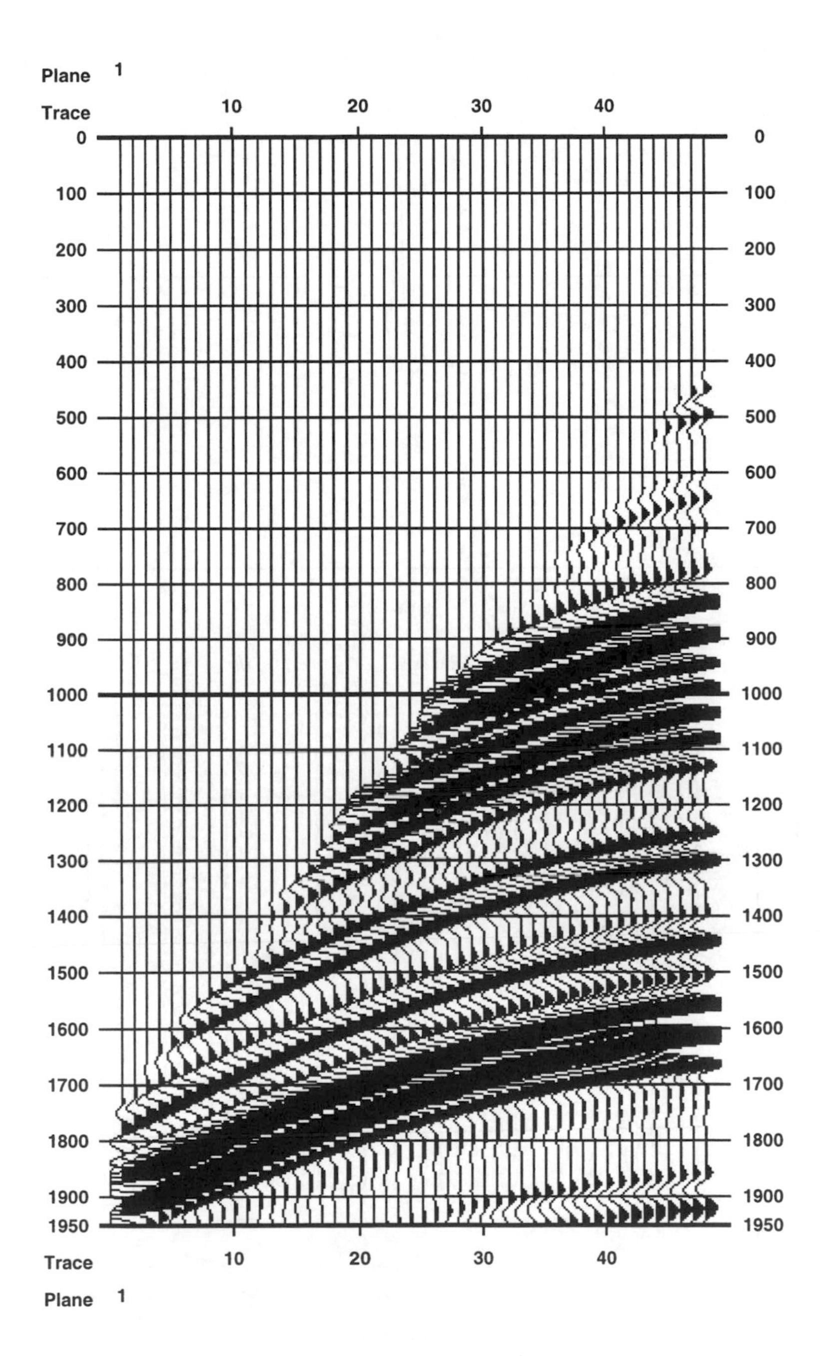

FIG. 4. *DIfference of data in Figures 2 and 3, plotted on same scale.*

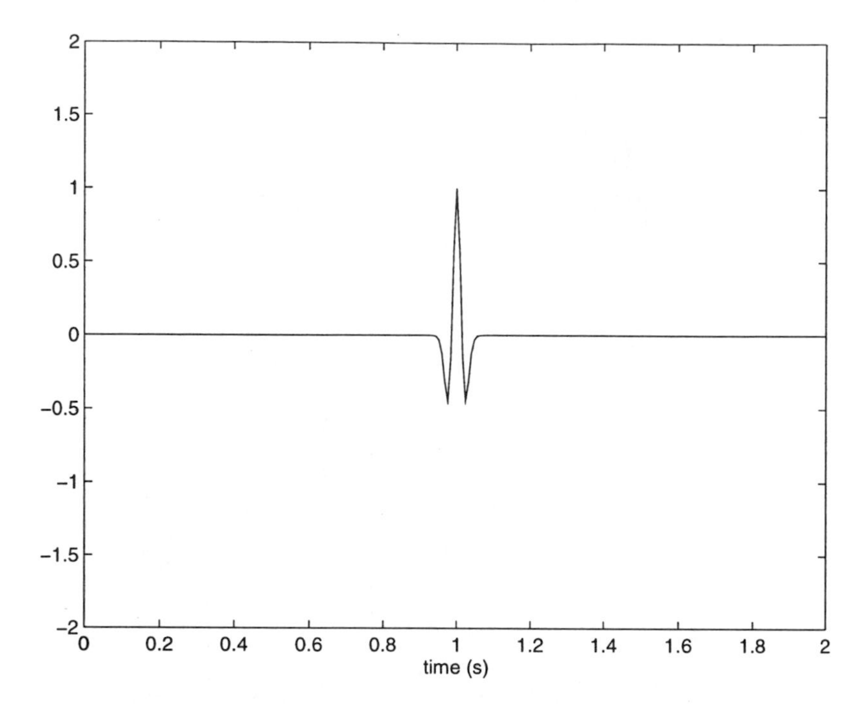

FIG. 5. *Ricker wavelet used to generate the data*

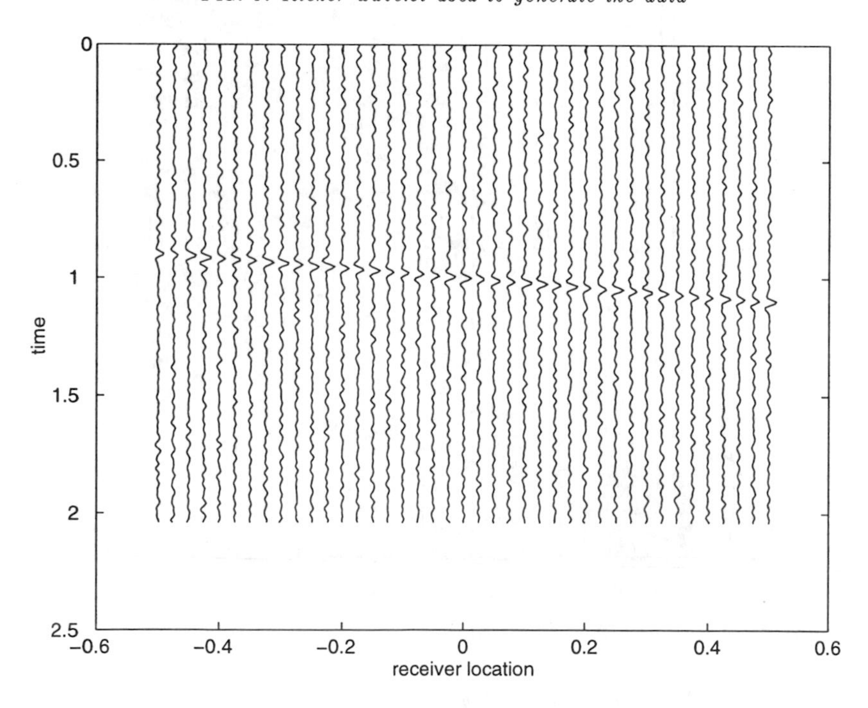

FIG. 6. *Synthetic plane wave data*

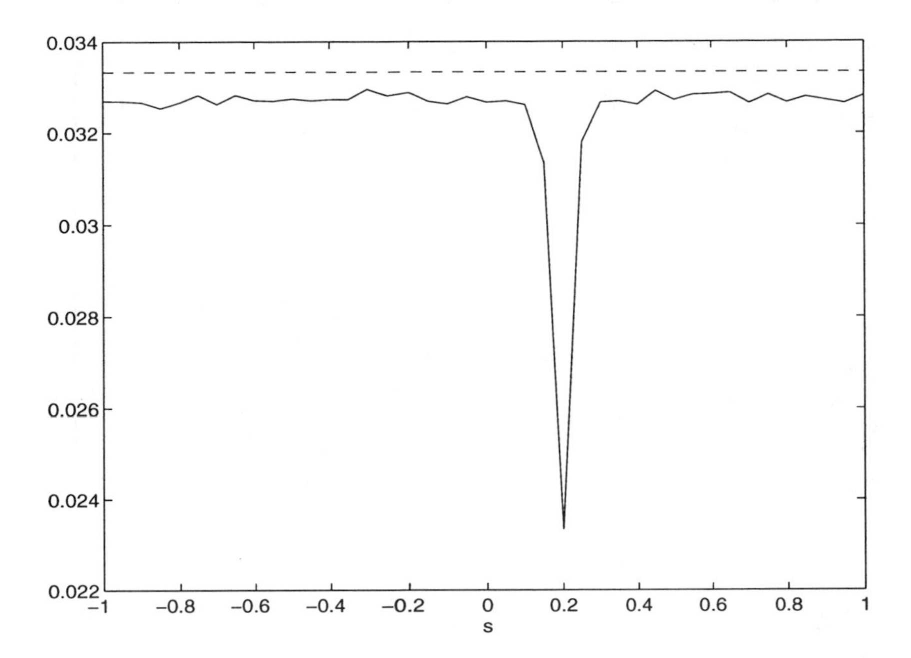

FIG. 7. *Output least-squares objective function; dashed line represents 100% misfit*

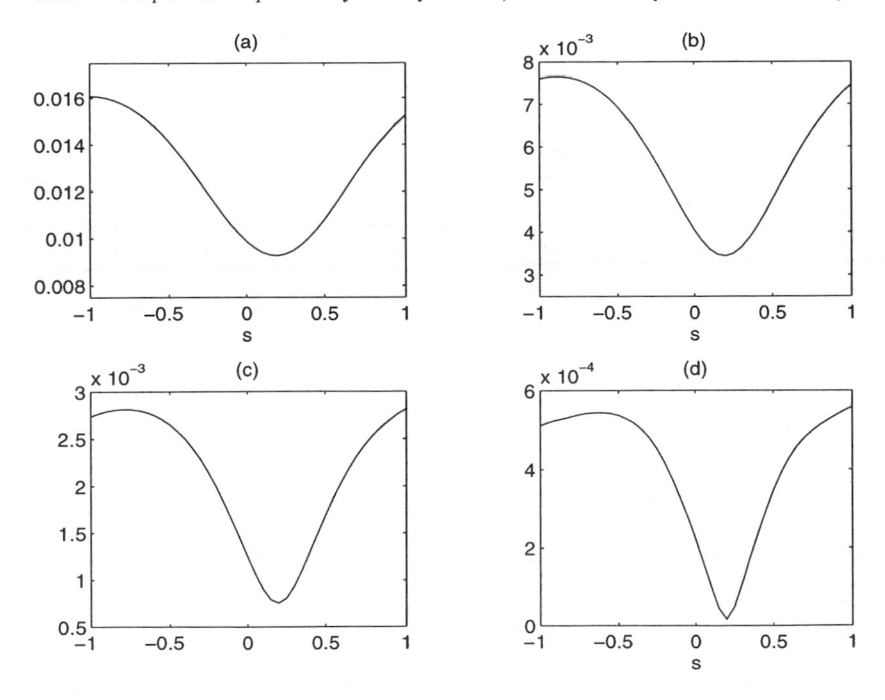

FIG. 8. J_ϵ *for (a)* $\epsilon = 0.05$, *(b)* $\epsilon = 0.1$, *(c)* $\epsilon = 0.15$, *and (d)* $\epsilon = 0.2$.

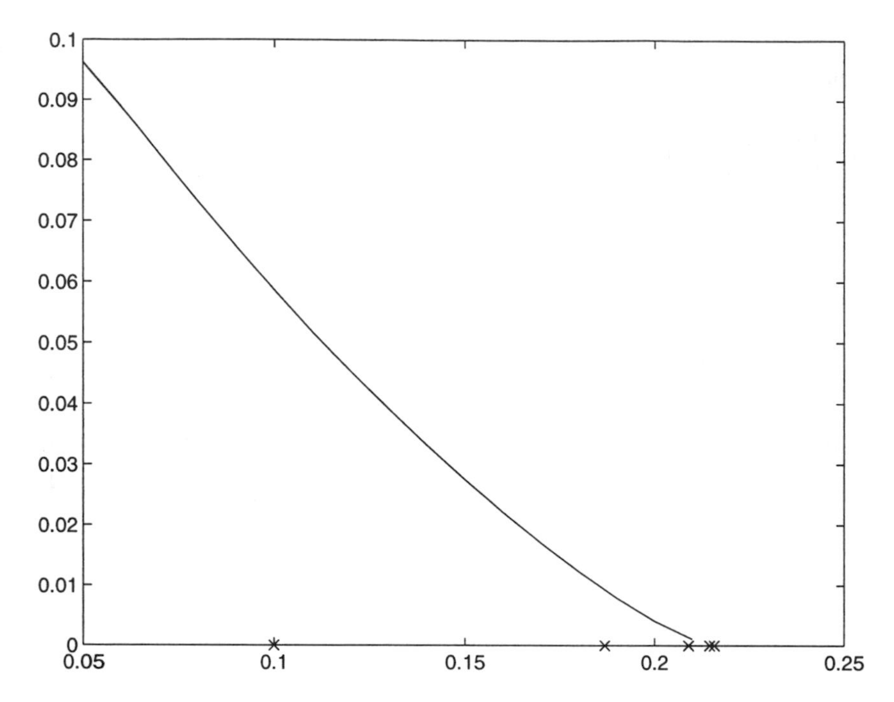

FIG. 9. *$\tilde{H}(\epsilon)$ versus ϵ; the progress of Newton's method for finding ϵ^* is indicated on the ϵ-axis*

REFERENCES

[Beylkin, 1985] Beylkin, G. (1985). Imaging of discontinuities in the inverse scattering problem by inversion of a causal generalized radon transform. *J. Math. Phys.*, 26:99–108.

[Dennis and Schnabel, 1983] Dennis, Jr., J. and Schnabel, R. (1983). *Numerical Methods for Unconstrained Optimization and Nonlinear Equations.* Prentice-Hall, Englewood Cliffs.

[Fiacco and McCormick, 1990] Fiacco, A.V. and McCormick, G.P. (1990). *Nonlinear programming: sequential unconstrained minimization techniques.* Classics in Applied Mathematics. Society for Industrial and Applied Mathematics, Philadelphia.

[Gauthier et al., 1986] Gauthier, O., Tarantola, A., and Virieux, J. (1986). Two-dimensional nonlinear inversion of seismic waveforms. *Geophysics*, 51:1387–1403.

[Gockenbach et al., 1995] Gockenbach, M., Symes, W., and Tapia, R. (1995). The dual regularization approach to seismic velocity inversion. *Inverse Problems*, 11(3):501–531.

[Hebden, 1973] Hebden, M.D. (1973). An algorithm for minimization using exact second derivatives. Technical Report TP515, A.E.R.E. Harwell.

[Kim and Symes, 1996] Kim, S. and Symes, W.W. (1996). Smooth detectors of linear phase. *Inverse Problems.* Submitted.

[Kolb et al., 1986] Kolb, P., Collino, F., and Lailly, P. (1986). Prestack inversion of a 1D medium. *Proceedings of IEEE 74*, pages 498–506.

[Mora, 1986] Mora, P. (1986). Nonlinear 2-d elastic inversion of multi-offset seismic data. *Geophysics*, 52:1211–1228.

[Moré, 1977] Moré, J.J. (1977). The Levenberg-Marquardt algorithm: implementation and theory. In Watson, G.A., editor, *Numerical Analysis*, Springer Verlag. Lecture Notes in Math. 630.

[Santosa and Symes, 1989] Santosa, F. and Symes, W. (1989). *An Analysis of Least-Squares Velocity Inversion*, volume 4 of *Geophysical Monographs*. Soc. of Expl. Geophys., Tulsa.

[Scales et al., 1991] Scales, J., Smith, M., and Fischer, T. (1991). Global optimization methods for highly nonlinear inverse problems. In Cohen, G., Halpern, L., and Joly, P., editors, *Mathematical and Numerical Aspects of Wave Propagation Phenomena*, pages 434–444. SIAM, Philadelphia.

[Sevink and Herman, 1994] Sevink, A.G.J. and Herman, G.C. (1994). Fast iterative solution of sparsely sampled seismic inverse problems. *Inverse Problems*, 10:937–948.

[Symes, 1991] Symes, W. (1991). The reflection inverse problem. In Cohen, Halpern, and Joly, editors, *Mathematical and Numerical Aspects of Wave Propagation Phenomena*, pages 423–433, SIAM, Philadelphia.

[Symes, 1994] Symes, W.W. (1994). The plane wave detection problem. *Inverse Problems*, 10:1361–1391.

[Virieux et al., 1992] Virieux, J., Jin, S., Madariaga, R., and Lambaré, G. (1992). Two dimensional asymptotic iterative elastic inversion. *Geophys. J. Int.*, 108:575–588.

PIECEWISE DIFFERENTIABLE MINIMIZATION FOR ILL-POSED INVERSE PROBLEMS

YUYING LI*

Abstract. Based on minimizing a piecewise differentiable l_p function subject to a single inequality constraint, this paper discusses algorithms for a discretized regularization problem for ill-posed inverse problems. We examine computational challenges of solving this regularization problem. Possible minimization algorithms such as the steepest descent method, iteratively weighted least squares (IRLS) method and a recent globally convergent affine scaling Newton approach are considered. Limitations and efficiency of these algorithms are demonstrated using the geophysical traveltime tomographic inversion and image restoration applications.

Key words. inverse problem, tomographic inversion, image restoration, total variation, steepest descent, projected gradient, IRLS, affine scaling and Newton.

1. Minimization and ill-posed inverse problems. Minimization algorithms have long been used in regulating an ill-posed inverse problem. Assuming that a desired property of a solution is known *a priori*, an ill-posed inverse problem can be regulated by solving a constrained minimization problem. In particular, properties expressed in nondifferentiable form have increasingly been found more appropriate in many applications. Discretization of such a regularization problem often leads to minimizing a large-scale piecewise differentiable function with a single constraint. In this paper, we consider regularization using piecewise differentiable minimization, possibly with a single inequality constraint.

Consider an ill-posed inverse problem,

$$\mathcal{A}u = f, \tag{1.1}$$

where $\mathcal{A}$ is an operator in a Hilbert space. Assume that $\|\cdot\|_2$ denotes the Euclidean norm and an *a priori* condition (e.g., continuity and boundedness) of the desired solution is given by $\|\mathcal{B}u\|_2 \leq \rho$ for some linear operator $\mathcal{B}$ and $\rho > 0$. Depending on applications, a reasonable solution to (1.1) can

---

* Department of Computer Science, Cornell University, Upson Hall, Ithaca, New York 14853-7501.

This paper is written for the proceedings of the workshop *Large-Scale Optimization, with Applications to Inverse Problems, Optimal Control and Design, & Molecular and Structural Optimization* in IMA, Minnesota, July 1995. Research is partially supported by the Applied Mathematical Sciences Research Program (KC-04-02) of the Office of Energy Research of the U.S. Department of Energy under grant DE-FG02-90ER25013.A000, and in part by NSF through grant DMS-9505155 and ONR through grant N00014-96-1-0050, and by the Cornell Theory Center which receives major funding from the National Science Foundation and IBM corporation, with additional support from New York State and members of its Corporate Research Institute.

be computed by solving either

$$\min_{u \in U} \|\mathcal{A}u - f\|_2 \quad \text{subject to} \quad \|\mathcal{B}u\|_2 \le \rho, \tag{1.2}$$

or

$$\min_{u \in U} \|\mathcal{B}u\|_2 \quad \text{subject to} \quad \|\mathcal{A}u - f\|_2 \le \delta. \tag{1.3}$$

The well-known constrained least squares approach, e.g., [14] for image restoration, corresponds to solving either (1.2) or (1.3).

Increasingly, properties involving a nonsmooth piecewise differentiable measurement have been found more appropriate in many applications [20,22,21,9,1,23,15,18,26]. We consider image restoration, for example. Let $u(x, y)$ denote the intensity of an image in a region Ω. Osher et al propose [20,22,21] to minimize total variation $\int_\Omega \sqrt{u_x^2 + u_y^2}\, dxdy$ to achieve image restoration:

$$\begin{aligned} &\min_u \int_\Omega \sqrt{u_x^2 + u_y^2}\, dxdy \\ &\text{subject to} \quad \int_\Omega ((\mathcal{A}u)(x,y) - u_0(x,y))^2\, dxdy = \sigma^2. \end{aligned} \tag{1.4}$$

This is a departure from the popular constrained least squares approach where the functional to be minimized is $\int_\Omega u_x^2 + u_y^2\, dxdy$. The total variation functional is not differentiable everywhere. Let U be a matrix representation of a 2-D image and u be a vector representation, a discretization of the problem (1.4) gives

$$\begin{aligned} &\min_{u \in \Re^n} \sum_l \|B_l u\|_2 = \sum_{ij} \sqrt{(U_{i,j+1} - U_{i,j})^2 + (U_{i+1,j} - U_{i,j})^2} \\ &\text{subject to} \quad \|Au - u_0\|_2 \le \sigma, \end{aligned} \tag{1.5}$$

where $B_l u$ is a discretized approximation to $[u_x; u_y]$ and Au is a discretized approximation to $\mathcal{A}u$. Applying the same total variation minimization principle, Li and Santosa [17] consider a slightly different discretized regularization problem:

$$\begin{aligned} &\min_{u \in \Re^n} \|Bu\|_1 = \sum_{ij} |U_{i,j+1} - U_{i,j}| + |U_{i+1,j} - U_{i,j}| \\ &\text{s.t.} \quad \|Au - u_0\|_2 \le \sigma. \end{aligned} \tag{1.6}$$

Similarly, a component of Bu denotes a discretization of u_x or u_y.

In geophysical traveltime tomographic inversion applications [23,15,18,26], subsurface slowness structure can be determined based on the inverse problem

$$t(ray) = \int_{ray} s(x, y) d\ell.$$

In [23], a discretized linear traveltime inversion problem is considered:

$$\min_{u \in \Re^n} \|Bu - c\|_1, \tag{1.7}$$

here B denotes a distance matrix, c is the recorded traveltime perturbation and $Bu = c$ is an overdetermined system. More generally, a linear l_p problem

$$\min_{u \in \Re^n} \|Bu - c\|_p, \tag{1.8}$$

has been considered [15,31].

The discretized regularization problems (1.7) and (1.8) are examples of a general problem formulation:

$$\begin{aligned} &\min_{u \in \Re^n} \phi(u) = \|Bu - c\|_p \\ &\text{subject to} \quad \|Au - f\|_2 \le \sigma, \qquad 1 \le p < 2, \end{aligned} \tag{1.9}$$

where A and B are matrices of the appropriate sizes and $\sigma > 0$ is a given parameter. Further generalization to a minimization problem using the more complicated objective function $\sum_{i=1}^{m} \|B_i u - c_i\|_p$ with the same inequality constraint of (1.10) (which then includes the formulation (1.5)) is worthy of future research.

Problem (1.10) is a piecewise differentiable minimization possibly with a single inequality constraint (if σ is infinity then the problem is unconstrained). Solving the minimization problem (1.10) is a computationally challenging task, particularly since a problem instance can be very large. However, if a reliable and efficient computational method can be developed to solve (1.10), then this minimization approach has great potential for solving various ill-posed inverse problems in many different applications.

The main objective of this paper is to compare a few possible algorithms for minimizing a piecewise differentiable function subject to a single quadratic inequality constraint. Image restoration and traveltime inversion applications are used to compare different computational algorithms. In §2, we illustrate the usefulness of the piecewise differentiable minimization for regulating ill-posed inverse problem using a traveltime tomographic inversion example. We analyze in §3 applicability of the steepest descent, iteratively reweighted least squares (IRLS), and a recent globally convergent affine scaling Newton approach for minimizing unconstrained piecewise differentiable l_p functions. We emphasize that the steepest descent method is not guaranteed to converge to a solution for a *nondifferentiable* minimization problem and can indeed fail in practice. In §4, the affine scaling Newton approach is adapted to further include the single inequality constraint. This algorithm is demonstrated with the image restoration application. Concluding remarks are given in §5.

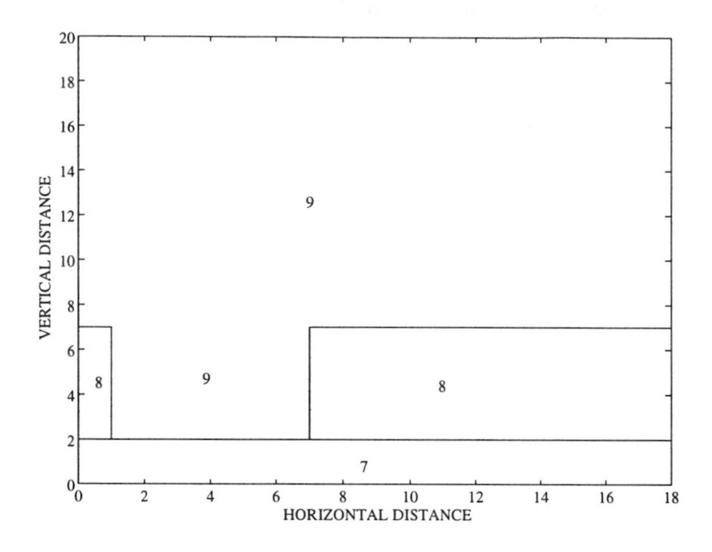

FIG. 1. *Geometry for Vertical Seismic Profile*

2. Piecewise differentiable measurements. Perhaps due to simplicity and availability of algorithms for smooth minimization, simple smooth convex function has been most widely used for data measurement. Increasingly, however, it has been recognized that a nondifferentiable measurement, e.g., a piecewise differentiable l_p norm, can be more appropriate in many different applications [25,15,26,20,22,21,9,1,17]. Computational algorithms have been developed for problems using these piecewise differentiable measurement [4,2,15,29,11,16,31,32,5]. Specifically, the piecewise linear l_1 measurement has become an increasingly attractive alternative to the 2-norm due to an appealing property of the l_1 solution: a small number of isolated large errors in the data typically do not change the solution of an unconstrained l_1 problem (e.g.,[10,23,26]).

To illustrate the usefulness of the piecewise differentiable minimization problem (1.10), we use a synthetic geophysical tomographic inversion problem described in [23], in which a square velocity anomaly is to be reconstructed.

Fig. 1 shows the geometry of this synthetic vertical seismic profile: there exists a square velocity anomaly in the middle layer. Fig. 2 shows the ray illumination from 18 sources on the surface to 18 receivers on the left side of the model. Straight rays are assumed and the lower left triangular region is covered with 136 square cells of constant slowness (reciprocal of velocity).

The following unconstrained piecewise differentiable minimization is

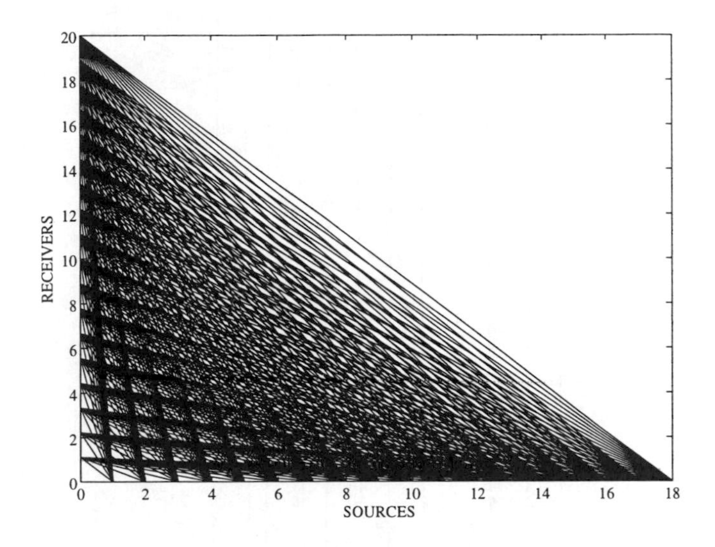

FIG. 2. *Ray Illumination from 18 Sources and 18 Receivers*

used in [15,23] for geophysical traveltime tomographic inversion

$$\min_{u \in \Re^n} \phi(u) = \|Bu - c\|_p, \quad 1 \leq p < 2,$$

where B is the distance matrix and c is the measured traveltime. If there is no error in c, there exists a solution satisfying $Bu = c$. This exact l_p solution is given in Fig. 3.

In general, however, errors exist in the measured traveltime c and different choices of the norm p lead to different solutions. To demonstrate this, we add both large spike errors and small Gaussian noise in the measured traveltime c, the least squares and l_1 reconstructions are shown in Fig. 4. Compared to Fig. 3, it is clear that the l_1 solution is less affected by the large spike error and thus is preferable in this context.

The l_p solution with $1 < p < 2$ has been used less often in practice (e.g, [23], [31]). For the synthetic seismic tomographic inversion problem considered, the l_p solutions, with p close to unity, have similar error resistance property, as illustrated in Fig. 6.

3. Minimization algorithms for unconstrained l_p problems. Solving a constrained piecewise differentiable problem (1.10) is challenging, especially when a problem is large (which is typical for many applications). We first focus on dealing with piecewise differentiability and consider

$$\min_{u \in \Re^n} \phi(u) = \|Bu - c\|_p, \qquad 1 \leq p < 2, \tag{3.1}$$

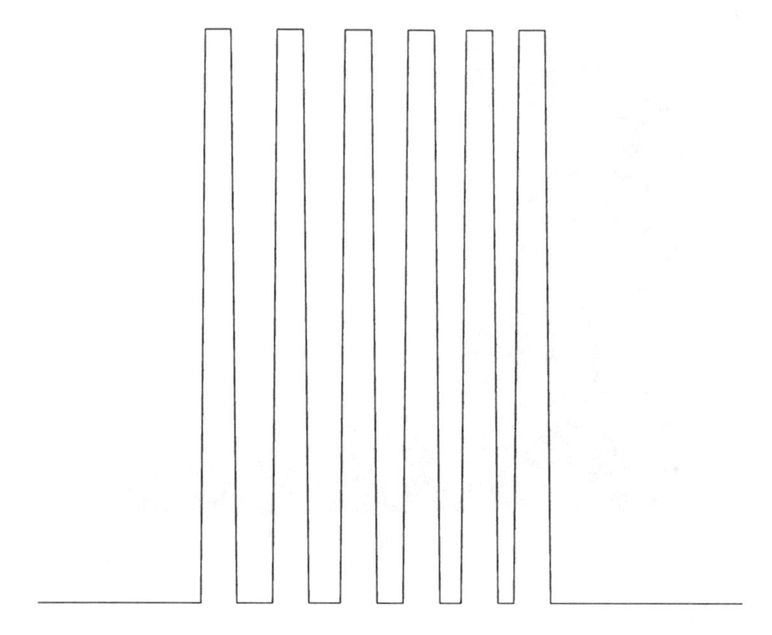

FIG. 3. *The Exact Solution for the Synthetic Geophysical Tomographic Inversion Example*

where $B \in \Re^{m \times n}$. We delay discussion on handling of the additional single inequality constraint until §4.

Let $r = Bu - c$ denote the residual vector. The objective function $\phi(u)$ in (3.1) is generally not differentiable everywhere, e.g., the gradient does not exist when a residual component $r_i = 0$. When $p = 1$, the linear l_1 problem can be solved by finite simplex algorithms based on linear programming [28] or a projected gradient method [4,3,6]. In this paper, we do not discuss this type of algorithms because it is less suitable for large-scale problems.

Most algorithms for unconstrained smooth nonlinear minimization problem are iteratively descent methods. At each iteration k, given a current approximation u^k to the solution, a descent direction d^k is computed and a new iterate $u^{k+1} = u^k + \alpha^k d^k$ is determined which produces a sufficient decrease of the objective function, i.e., $\phi(u^{k+1})$ is significantly less than $\phi(u^k)$ (The stepsize α^k can be an approximation to the minimizer of $\phi(u^k + \alpha d^k)$). This minimization process can be terminated when decrease in function $\phi(u)$ is sufficiently small, e.g.,

$$\frac{\phi(u^k) - \phi(u^{k+1})}{\phi(u^{k+1})} \leq tol \quad \textbf{or} \quad k > itmax,$$

where tol is a small positive number denoting acceptable accuracy and $itmax$ is the maximum number of iterations allowed. In this paper, we consider three types of descent directions including the simple steepest

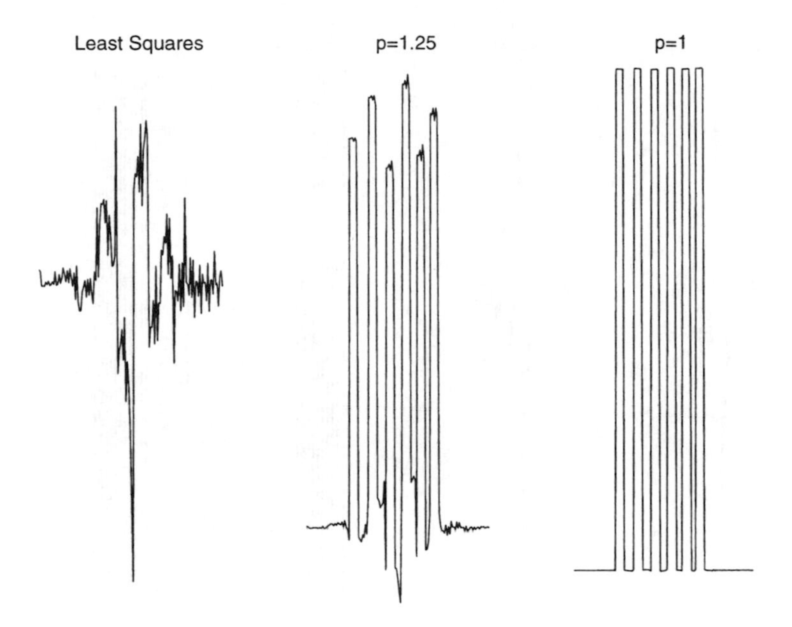

FIG. 4. *Comparison of Different Choice of Norms*

descent, iteratively reweighted least squares, and the more sophisticated affine scaling Newton directions. Subsequently we demonstrate the effectiveness of each descent direction in the context of the linear l_p problem (3.1).

3.1. Steepest descent directions. If $\phi(u)$ is differentiable at u^k, then $d^k = -\nabla\phi(u^k)$ is a descent direction. The most appealing aspect of steepest descent is its low cost. For example, when $p = 1$, the gradient $\nabla\phi(u^k) = B^T \text{sgn}(Bu^k - c)$ (when exists) can be computed with a matrix vector product. (In this paper, we define $\text{sgn}(0) = 1$.) If $\phi(u)$ is not differentiable everywhere, e.g., for $\phi(u) = \|Bu - c\|_1$, then the steepest descent method does not, in general, lead to convergence to a solution of the problem (3.1). Moreover, even if $\{u^k\}$ converges to a solution, the convergence can be extremely slow.

To illustrate this, the reconstruction of the velocity for the traveltime inversion example in Fig. 1, using steepest descent, is given in Fig. 5. The descent method using steepest descent directions fails to provide any visible improvement over the starting least squares reconstruction after 500 iterations (maximum number of iterations allowed). Moreover, convergence will not be achieved eventually.

Another fairly common approach for overcoming nondifferentiability, used in many applications, is smoothing. Smoothing eliminates nondifferentiability by slightly perturbing the objective function. For example, one may choose p close to one in (3.1) in the hope of using minimization

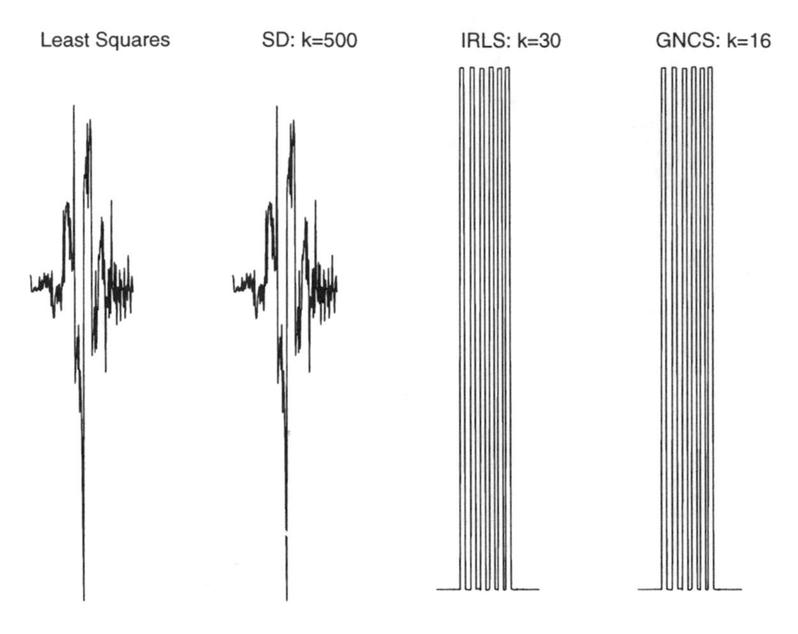

FIG. 5. *Comparisons of Different Methods When* $p = 1$

techniques for a smooth problem, e.g., the steepest descent method. We believe that this is a wrong reason for choosing p. In particular, the steepest descent method can fail for the slightly perturbed problem, similar to the case when applied directly to the original nondifferentiable problem. To illustrate this, we apply the steepest descent method for minimizing $||Bu - c||_p$, with $p = 1.001$. As illustrated in Fig. 6, the steepest descent method again is unable to provide significant improvement over the least squares reconstruction after 500 iterations, even though the solution of (3.1) with $p = 1.001$ is essentially the same as the exact solution in Fig. 3 (the solution is computed using the affine scaling Newton directions described in §3.3).

3.2. Iteratively reweighted least squares. In order to compute a solution with improved efficiency, an iteratively reweighted least squares (IRLS) approach has often been used for the linear l_p problem (3.1), e.g., [24,7,19]. The IRLS algorithm has become popular due to its simplicity: a weighted least squares problem is solved at the iteration k to get a descent direction and hence an improved approximation u^{k+1}. For some applications, e.g., [8], this algorithm can indeed be useful, particularly if an accurate solution is not necessary. However, there is evidence [19] which does not recommend this method in general due to its linear convergence.

The IRLS algorithm can be derived by considering a Newton step for $\phi(u)$ but ignoring the fact that the objective function may not be sufficiently smooth. Let r denote the residual vector $Bu - c$. Consider u^k and assume

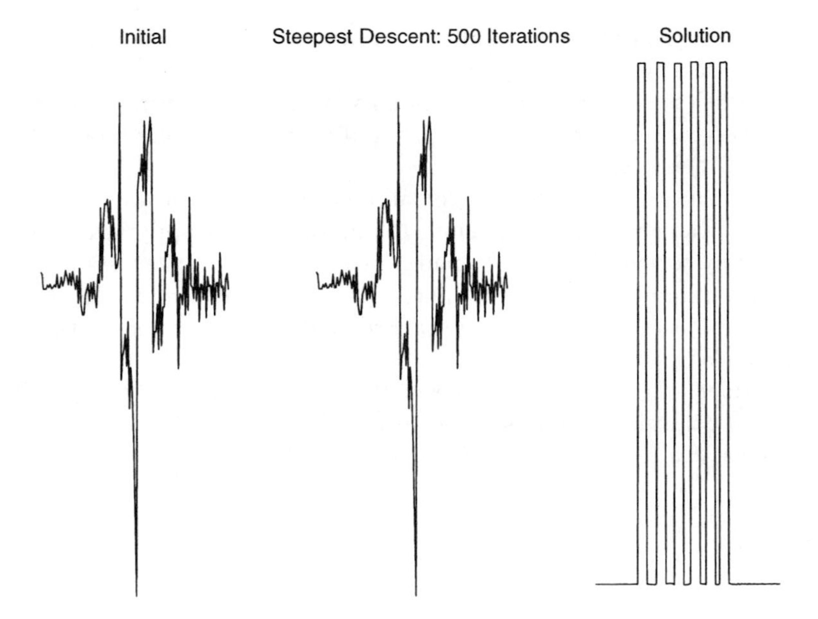

FIG. 6. *Reconstruction Using Steepest Descent with Smoothing ($p = 1.001$)*

that the corresponding $|r^k| > 0$. A Newton step for minimizing $\phi(u) = ||Bu - c||_p$ is

$$B^T \operatorname{diag}(|r^k|^{p-2})Bd = -\frac{1}{p(p-1)} \nabla\phi(u^k),$$

which is the normal equation for the weighted least squares problem

$$\operatorname{diag}(|r^k|^{\frac{p-2}{2}})Bd \stackrel{\text{LS}}{=} -\frac{1}{p(p-1)} \operatorname{diag}(|r^k|^{\frac{2-p}{2}})g^k. \tag{3.2}$$

The direction d^k computed from above equations is descent, and, the iterates $\{u^k\}$ generated by IRLS satisfy $u^{k+1} = u^k + (p-1)d^k$ and converge linearly to a solution [30,19]. A line search acceleration of IRLS can be easily made to achieve a better reduction of $\phi(u)$ along the descent direction d^k. Subsequently, when we refer to the IRLS method, we are referring to this accelerated IRLS.

The reconstruction using the IRLS method is given in Fig. 5. The iterates from the IRLS indeed converge in 30 iterations and a good reconstruction is achieved.

3.3. A globally convergent affine scaling newton approach. The IRLS method successfully achieves convergence with a significant cost: solving a weighted least squares problem at each iteration. With increasing computing power and availability of reliable and efficient least squares solver on sequential and parallel computers, the additional cost of IRLS may become more acceptable.

Nonetheless, the IRLS method is only a linearly convergent method. Fast convergence is essential for efficiency if an accurate solution of a large problem (3.1) is desired.

In [16,12], a globally convergent affine scaling Newton method is developed for the problem (3.1). This globalized Newton approach (GNCS) is derived by considering complementarity conditions for the piecewise differentiable minimization (3.1). In particular, a descent direction is defined based on both the current primal iterate u^k and approximations to the dual multipliers λ^k.

Consider $\min_x \|Bu - c\|_p$. Let $r = Bu - c$ and $g = p|r|^{p-1}\text{sgn}(r)$. The optimality conditions for this problem are described in both x and dual multipliers λ: a point u is a solution to (3.1) if there exists λ such that

$$\begin{aligned} &\text{diag}(r)(g - \lambda) = 0 \\ &B^T\lambda = 0, \quad Zr = Zb. \end{aligned} \tag{3.3}$$

with $|\lambda| \leq |g|$. Here Z is a matrix whose rows form a basis for the null space of B, i.e., $Z^T = \text{null}(B)$. We note that *explicit knowledge of Z is not necessary in the actual algorithm.*

Let $D_r = \text{diag}(|r|)$ and $D_\lambda = \text{diag}(\text{sgn}(r))\text{diag}(pg - \lambda)$. A Newton step $d \in \Re^n$ and $d_w \in \Re^{m-n}$ for (3.3) can be computed:

$$[D_\lambda^k B, -D_r^k Z^T] \begin{bmatrix} d \\ d_w \end{bmatrix} = -D_r^k(g^k - Z^T w^k). \tag{3.4}$$

Observing the special structure of the linear system (3.4), it is easy to see that the Newton direction d^k equals

$$d^k = -(B^T D_\lambda^k {D_r^k}^{-1} B)^{-1} B^T g^k. \tag{3.5}$$

Globalization of the Newton step is typically required in the minimization context to ensure proper descent directions are defined everywhere. Let $\theta^k \in \Re^m$ denote the optimality measurement below

$$\begin{aligned} &\theta = \tfrac{\eta e}{\gamma|g| + \eta e}, \\ &\eta = \max(\max(\tfrac{|D_r(g-\lambda)|}{\phi(r_0)}), \max(\max(|\lambda| - |g|, 0))) \end{aligned} \tag{3.6}$$

where $0 < \gamma < 1$ is a constant and e is the vector of all ones. We use the Matlab [27] definition for the function max: $\max(x)$ denotes the maximum component of a vector x and the value of $\max(x, y)$ is a vector whose components are the maximum of the corresponding components of x and y.

There are many possible ways to globalize. An example which works well computationally is given in [12,16]. This globalization replaces the diagonal matrix D_λ^k by a combination D_θ^k of diag(g^k) and D_λ^k using the optimality measurement $\theta^k > 0$:

$$D_\theta^k = \mathrm{diag}(|\theta^k g^k + (e - \theta^k)(pg^k - \lambda^k)|).$$

Hence a globalized descent direction d^k is computed from

$$d^k = -B(B^T D_\theta^k {D_r^k}^{-1} B)^{-1} B^T g^k. \tag{3.7}$$

It is easy to see that this direction d^k can be computed as a weighted least squares solution

$$(D_\theta^k {D_r^k}^{-1})^{\frac{1}{2}} Bd \overset{\text{LS}}{=} -(D_\theta^k {D_r^k}^{-1})^{-\frac{1}{2}} g^k. \tag{3.8}$$

Dual multipliers approximations can be updated:

$$\lambda^{k+1} = D_\theta^k {D_r^k}^{-1} Bd^k + g^k.$$

The new iterate is

$$u^{k+1} = u^k + \alpha^k d^k,$$

where α^k is the stepsize which approximately minimizes the piecewise differentiable function $\phi(u^k + \alpha d^k)$ and can be determined easily (we refer a reader to [12,16] for details of the line search).

We observe that the weighted least squares problems (3.2) and (3.8) have the same coefficient matrix B but with different diagonal scalings. Hence the direction d^k (3.8) can be computed with the same cost as that of the IRLS method (3.2). The approximate multipliers can be obtained with negligible cost. Fig. 5 illustrates the reconstruction for the same example in Fig. 1 using the described affine scaling Newton method. This affine scaling Newton method takes roughly half the number of iterations of the IRLS method.

We have performed many computational experiments which indicate that the globally convergent affine scaling Newton approach [12,16] generally takes significantly fewer iterations than the IRLS method (less than half). Table 3.3 illustrates typical behaviors of the two methods in iteration accounts for some randomly generated problems (3.1) with $p = 1$. The stopping tolerance is 10^{-12} and the maximum number of iterations allowed is 50.

4. Solving a linear l_p problem with a single quadratic constraint. For many ill-posed inverse problems, e.g., image restoration, an *a priori* condition on the desired solution is needed for regularization. Hence it may be appropriate to solve a constrained piecewise differentiable minimization problem (1.10). In this section, we will use image restoration as an example to illustrate how to incorporate the constraint information

Number of Steps	$m = 200$	
n	GNCS	IRLS
10	17	50
30	17	50
50	15	50
70	21	50
90	15	50
110	14	50
130	17	50
150	13	50
170	13	50
190	9	50

using the affine scaling Newton method [12] described in §3.3. Specifically, we consider solving (1.7):

$$\begin{aligned} &\min_{u \in \Re^n} \|Bu - c\|_1 \\ \text{subject to} \quad &\|Au - u_0\|_2 \leq \sigma. \end{aligned} \tag{4.1}$$

It is well recognized that generating iterates to follow nonlinear constraints is difficult; even a single nonlinear constraint adds substantial computational difficulty. The classical approaches for nonlinear constraints include the l_1 penalty and Lagrangian method [13]. The difficulty with these methods is determination of the *a priori* unknown penalty parameter. We believe that a more appealing approach here is to maintain feasibility for the single constraint $\|Ax - u_0\|_2 \leq \sigma$ explicitly.

For image restoration application, a solution typically appears on the quadratic surface. If the solution lies strictly inside the constraint $\|Ax - u_0\|_2 < \sigma$, then the problem is essentially unconstrained and can be solved as described in §3.3. In general, a decision can be made as to whether the solution is exactly on the nonlinear constraint surface with available dual multipliers approximation. Employing the active set strategy on the single quadratic constraint, descent directions can be computed to leave or remain on this single constraint surface, e.g. [13]. We assume subsequently that the solution lies on the quadratic constraint surface $\|Au - u_0\|_2 = \sigma$.

Feasibility to the constraint $\|Au - u_0\|_2 \leq \sigma$ can be achieved by applying a minimization process (e.g., a conjugate gradient method) to the convex quadratic function $\|Au - u_0\|_2^2$ until feasibility $\|Au - u_0\|_2 \leq \sigma$ is obtained. This feasibility can be easily maintained using a line search with a correction [13,17]. We subsequently concentrate on computing a good descent direction for (4.2).

Let $J = [A^T(Au - u_0), B^T]$, $\hat{g} = [0; \text{sgn}(Bu - c)]$ and $\hat{r} = [\|A^T u - u_0\|^2 - \sigma^2; Bu - c]$. The nonlinear system which captures optimality of the constrained piecewise linear minimization problem (4.2) is

$$\begin{aligned} J\lambda &= 0, \\ \operatorname{diag}(\hat{r})(\hat{g} - \hat{\lambda}) &= 0. \end{aligned} \tag{4.2}$$

Note that the additional condition, $|\lambda_i| \leq 1$ for any $\hat{r}_i = 0$ and $i > 1$, needs to hold.

Let $\hat{D}_\lambda^k = \operatorname{diag}(\operatorname{sgn}(\hat{r}^k))\operatorname{diag}(\hat{g}^k - \hat{\lambda}^k)$. A Newton step for (4.2) is

$$\begin{bmatrix} \hat{\lambda}_1^k A^T A & J^k \\ \hat{D}_\lambda^k J^{k^T} & -\operatorname{diag}(\hat{r}^k) \end{bmatrix} \begin{bmatrix} d^k \\ \hat{\lambda}^{k+1} \end{bmatrix} = - \begin{bmatrix} 0 \\ \operatorname{diag}(\hat{r}^k)\hat{g}^k \end{bmatrix}. \tag{4.3}$$

Similar to discussion in §3.3, globalization is needed to ensure that descent directions are properly determined and the globalized Newton directions converge to the Newton direction (4.3) asymptotically. We consider a similar globalization as described before: we replace $\hat{D}_\lambda^k$ by

$$\hat{D}_\theta^k = \operatorname{diag}(|\hat{\theta}^k \hat{g}^k + (e - \hat{\theta}^k)(\hat{g}^k - \hat{\lambda}^k)|)$$

and consider a globalized Newton direction as below:

$$\begin{bmatrix} |\hat{\lambda}_1^k| A^T A & J^k \\ D_\theta^k J^{k^T} & -\operatorname{diag}(|\hat{r}^k|) \end{bmatrix} \begin{bmatrix} d^k \\ \hat{\lambda}^{k+1} \end{bmatrix} = - \begin{bmatrix} 0 \\ \operatorname{diag}(|\hat{r}^k|)\hat{g}^k \end{bmatrix}.$$

Eliminating the dual variables, we have

$$(|\hat{\lambda}_1^k| A^T A + J^k \hat{D}_\theta^k {\hat{D}_r^k}^{-1} J^{k^T}) d = -J^k \hat{g}^k.$$

Compared to (3.7), (4.4) has an additional term $|\hat{\lambda}_1^k| A^T A$ which represents the constraint curvature. Let $\hat{D}^k$ denote the diagonal scaling matrix below

$$\hat{D}^k = \begin{bmatrix} (\hat{D}_\theta^k {\hat{D}_r^k}^{-1})^{\frac{1}{2}} & 0 \\ 0 & \sqrt{|\hat{\lambda}_1^k| I} \end{bmatrix}. \tag{4.4}$$

A simple algebraic manipulation shows that the globalized direction d^k (3.8) can be computed from a weighted least squares solve

$$\hat{D}^k \begin{bmatrix} J^{k^T} \\ A \end{bmatrix} d^k \stackrel{\text{LS}}{=} -{\hat{D}^k}^{-1} \begin{bmatrix} \hat{g}^k \\ 0 \end{bmatrix}, \tag{4.5}$$

and

$$\hat{\lambda}^{k+1} = (\hat{D}_\theta^k {\hat{D}_r^k}^{-1}) d^k + \hat{g}^k.$$

A simple correction technique for maintaining feasibility can be incorporated to stay on the quadratic surface [17]. For efficient computation, the

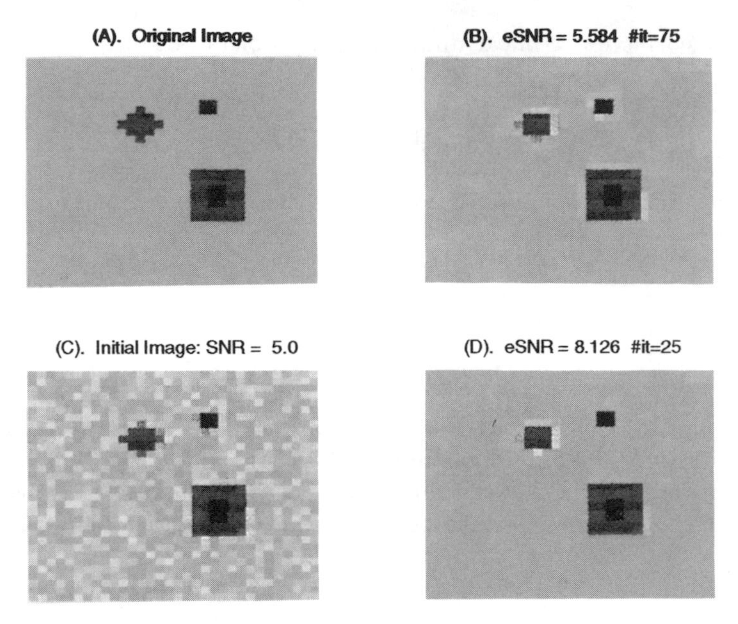

FIG. 7. *Comparisons of Affine Scaling Newton Directions and Steepest Descent Directions*

single dense row structure for the least squares problem (4.5) needs to be exploited.

To demonstrate the descent algorithm using the affine scaling Newton direction (4.5) for the constrained minimization problem (1.7), we generate the blurred and noisy image u_0 by a motion blurring function A, and adding a measured amount of random noise ε, i.e.,

$$u_0 = Au_{true} + \varepsilon.$$

The amount of noise in the noisy and blurred image u_0 is described by the signal-to-noise ratio (SNR):

$$\text{SNR} = 10 \log \frac{\text{variance of the blurred image}}{\text{variance of the noise}} \quad (dB).$$

When assessing the quality of the restored images, we consider the signal-to-noise ratio improvement e_{SNR}

$$e_{\text{SNR}} = 10 \log \frac{\|u_{true} - u_0\|_2^2}{\|u_k - u_{true}\|_2^2} \quad (dB).$$

When SNR is high, a larger signal-to-noise ratio improvement e_{SNR} suggests a better restoration.

In Fig. 7 and 8, we demonstrate the quality of image restoration for a 32-by-32 and 64-by-64 blurred and noisy images respectively. We apply

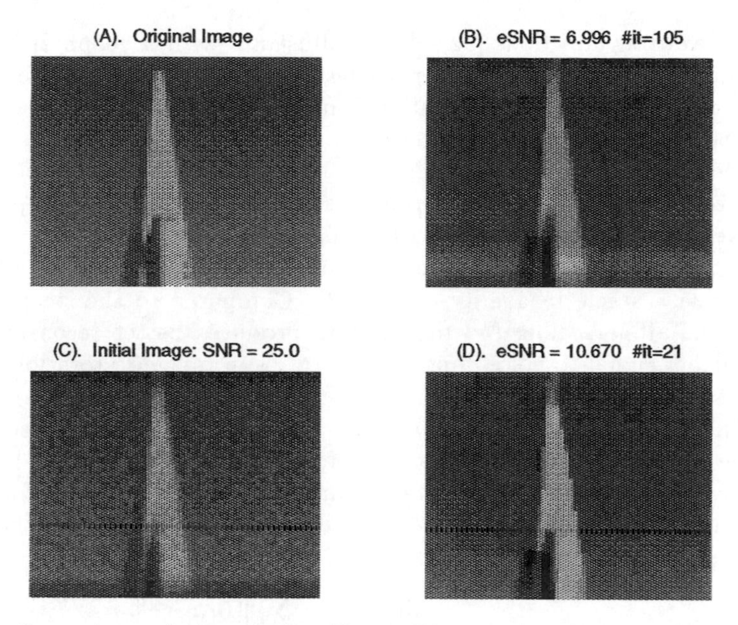

FIG. 8. *Comparisons of Affine Scaling Newton Directions and Steepest Descent Directions*

both the steepest descent and affine scaling Newton directions for comparison. The stopping criteria for these images are 0.5×10^{-4} with the maximum of 200 iterations allowed. The parameter $\sigma = 0.95 \times ||\varepsilon||_2$ where ε is the random white noise.

Unlike the geophysic traveltime inversion problem, surprisingly good improvements are achieved using steepest descent directions. However, the images computed using the affine scaling Newton directions are consistently better and achieve higher signal-to-noise ratio improvements. Compared to the affine scaling Newton direction, a steepest descent direction is significantly cheaper to compute. Nonetheless, the steepest descent method is more sequential than the Newton approach. In a parallel computing environment with a good parallel least squares solver, the affine scaling Newton method with a dominant work on the least squares solve may prevail if a sparse matrix A is available.

If the matrix A is not available or is very dense, the computation of an affine scaling Newton step (4.5) may be impossible or extremely expensive. We are currently investigating possibility of computing an affine scaling Newton step (4.5) iteratively.

5. Concluding remarks. Piecewise differentiable minimization with a single quadratic constraint has great potential in regulating ill-posed inverse problems. We analyze three descent methods using steepest descent, iteratively reweighted least squares and the more recent affine scaling Newton directions [12,16]. We illustrate with a geophysic travel-time inversion example, that the steepest descent can fail to achieve convergence for a piecewise differentiable minimization and produce entirely unacceptable results. The alternative iteratively reweighted least squares method is more computational extensive and can be inefficient due to its linear convergence. The globally convergent affine scaling Newton approach achieves quadratic convergence with negligible additional cost, compared to the IRLS approach. The affine scaling Newton approach is further adapted to include a single inequality constraint. Compared to the steepest descent, the affine scaling Newton method produces better reconstruction with higher signal-to-noise improvement ratios in image restoration applications.

The future investigation includes possible iterative solvers for computing the Newton steps for the globalized Newton approach [12,16]. The issue of preconditioner becomes extremely important in this regard. Moreover, algorithms for a more general piecewise differentiable problem of the form

$$
\begin{aligned}
&\min_{u \in \Re^n} \phi(u) = \sum_{i=1}^{t} \|B_i u - c\|_p \\
&\text{subject to} \quad \|Au - f\|_2 \le \sigma,
\end{aligned}
\tag{5.1}
$$

may be useful for various applications.

Acknowledgements. The author would like to thank Fadil Santosa for introducing image restoration problem to her, Tom Coleman for his comments, and Chunguang Sun for use of his sparse least squares solve routines.

REFERENCES

[1] S. Alliney and S. Ruzisky, *An algorithm for the minimization of mixed l_1 and l_2 norms with applications to Bayesian estimation*, IEEE Trans. on Signal Processing, 42 (1994), pp. 618–627.

[2] I. Barrodale and C. Phillips, *An improved algorithm for discrete Chebychev linear approximation*, in Proc. 4th Manitoba Conf. on Numer. Math., U. of Manitoba, Winnipeg, Canada (1974), pp. 177–190.

[3] I. Barrodale and F. Roberts, *An improved algorithm for discrete l_1 linear approximation*, SIAM J. Num. Anal., 10 (1973), pp. 839–848.

[4] I. Barrodale and A. Young, *Algorithms for best l_1 and l_∞ linear approximations on a discrete set*, Numer. Math., 8 (1966), pp. 256–306.

[5] R.H. Bartels, A.R. Conn, and Y. Li, *Primal methods are better than dual methods for solving overdetermined linear systems in the l_∞ sense?*, SIAM J. Numer. Anal., 26 (1989), pp. 693–726.

[6] R. H. Bartels, A. R. Conn, and J. W. Sinclair, *Minimization techniques for piecewise differentiable functions: the l_1 solution to an overdetermined linear system*, Siam J. Numer. Anal., 15 (1978), pp. 224–240.

[7] A.E. Beaton and J.W. Tukey, *The fitting of power series, meaning polynomials, illustrated on band-spectrographic data*, Technometrics, 16 (1974), pp. 147–185.

[8] P.J. BESL, J.B. BIRCH, AND L.T. WATSON, *Robust window operators*, Machine Vision and Applications, 2 (1989), pp. 179–191.

[9] C. BOUMAN AND K. SAUER, *A generalized Gaussian model for edge preserving MAP estimation*, IEEE Trans. on Image Processing, 2 (1993), pp. 298–310.

[10] J.F. CLAERBOUT AND F. MUIR, *Robust modeling with erratic data*, Geophysics, 38 (1973), pp. 826–844.

[11] T.F. COLEMAN AND Y. LI, *A global and quadratically-convergent method for linear l_∞ problems*, SIAM Journal on Scientific and Statistical Computing, 29 (1992), pp. 1166–1186.

[12] ———, *A globally and quadratically convergent affine scaling method for linear l_1 problems*, Mathematical Programming, 56 (1992), pp. 189–222.

[13] R. FLETCHER, *Practical Methods of Optimization: Volume 2, Constrained Optimization*, John Wiley and Sons (1981).

[14] B.R. HUNT, *The application of constrained least squares estimation to image restoration by digital computer*, IEEE Transactions on Computers, C-22 (1973), pp. 805–812.

[15] A.G. JOHN A. SCALES AND S. TREITEL, *Fast l_p solution of large, sparse, linear systems and application to seismic travel time tomography*, Journal of Computational Physics, 75 (1988), pp. 314–333.

[16] Y. LI, *A globally convergent method for l_p problems*, SIAM Journal on Optimization, 3 (1993), pp. 609–629.

[17] Y. LI AND F. SANTOSA, *An affine scaling algorithm for minimizing total variation in image enhancement*, IEEE Transactions on Image Processing, 5 (1996), pp. 987–995.

[18] L.R. LINES, *Applications of seismic traveltime tomography–a review*, in Geophysical Inversion (1992), pp. 333–337. SIAM.

[19] M.R. OSBORNE, *Finite Algorithms in Optimization and Data Analysis*, Wiley Series in Probability and Mathematical Statistics (1985).

[20] L. RUDIN AND S. OSHER, *Feature-oriented image enhancement using shock filters*, SIAM J. Numer. Anal., 27 (1990), pp. 919–940.

[21] L. RUDIN, S. OSHER, AND C. FU, *Total variation based restoration of noisy and blurred images*, SIAM J. Numer. Anal., to appear.

[22] L. RUDIN, S. OSHER, AND E. FATEMI, *Nonlinear total variation based noise removal algorithms*, Physics D., 60 (1992), pp. 259–268.

[23] J.A. SCALES, *Tomographic inversion via the conjugate gradient method*, Geophysics, 52 (1987), pp. 179–185.

[24] E.J. SCHLOMACHER, *An iterative technique for absolute deviations curve fitting*, J. Am. Stat. Assoc., 68 (1973), pp. 857–865.

[25] H. SPÄTH, *Mathematical Algorithms for Linear Regression*, Academic Press (1992).

[26] H.L. TAYLOR, S.C. BANKS, AND J.F. MCCOY, *Deconvolution with the l_1 norm*, Geophysics (1979), pp. 39–52.

[27] THE MATHWORKS INC., *Matlab Reference guide*, The MathWorks, Natick, Mass (1992).

[28] H.M. WAGNER, *Linear programming techniques for regression analysis*, J. Am. Stat. Assoc., 54 (1959), pp. 206–212.

[29] G.A. WATSON, *On two methods for discrete l_p approximation*, Computing, 18 (1977), pp. 263–266.

[30] J.M. WOLFE, *On the convergence of an algorithm for discrete l_p approximation*, Numer. Math., 32 (1979), pp. 439–459.

[31] R. YARLAGADDA, J.B. BEDNAR, AND T.L. WATT, *Fast algorithms for l_p deconvolution*, IEEE Transactions on Acoustics, Speech and Signal Processing, ASSP-33 (1985), pp. 174–182.

[32] Y. ZHANG, *A primal-dual interior point approach for computing the l_1 and l_∞ solutions of overdetermined linear systems*, Tech. Rep. Technical Report, Department of Mathematics and Statistics, University of Maryland (1991).

THE USE OF OPTIMIZATION IN THE RECONSTRUCTION OF OBSTACLES FROM ACOUSTIC OR ELECTROMAGNETIC SCATTERING DATA

PIERLUIGI MAPONI*, MARIA CRISTINA RECCHIONI†, AND FRANCESCO ZIRILLI‡

Abstract. We consider some three dimensional time harmonic acoustic and electromagnetic scattering problems for bounded simply connected obstacles. We consider the following inverse problem: from the knowledge of several far field patterns generated by the obstacle when hit by known incoming waves and from the knowledge of some a-priori information about the obstacle, i.e. boundary impedance, shape symmetry, etc., reconstruct the shape or the shape and the impedance of the obstacle. There are a large number of effective numerical methods to solve the direct problem associated with this inverse problem, but techniques to solve the inverse problem are still in their infancy. We reformulate the inverse problem as two different unconstrained optimization problems. We present a review of results obtained by the authors on the inverse problem and we give some ideas concerning the solution of the direct problem by efficient parallel algorithms.

1. Introduction. Let $\mathbf{R}^3$ be three dimensional real Euclidean space, $\underline{x} = (x_1, x_2, x_3)^T \in \mathbf{R}^3$ be a generic vector where the superscript T denotes transpose, and $(\cdot,\cdot)$ denotes the Euclidean scalar product, $\|\cdot\|$ the Euclidean norm and $[\cdot,\cdot]$ the vector product. Occasionally we allow complex vectors to be the argument of real Euclidean scalar or vector products.

Let $D \subset \mathbf{R}^3$ be a bounded simply connected domain, with smooth boundary ∂D. We consider D as a scatterer contained in a medium filling $\mathbf{R}^3 \backslash D$ where the acoustic or electromagnetic waves are propagating. In the following we assume that the origin of the coordinate system lies inside D.

The acoustic scattering problems. We consider $\mathbf{R}^3 \backslash D$ filled by a uniform nonviscous fluid in thermodynamic equilibrium with zero heat conductivity, adiabatic compressibility κ, temperature T, pressure P and mass density per unit volume ρ. When an incoming acoustic wave is propagated in this medium the pressure at a point $\underline{x} \in \mathbf{R}^3 \setminus \overline{D}$ at a time $t > 0$ is $P + p^i(\underline{x}, t)$, where p^i is the acoustic pressure of the wave. If we assume that p^i is small compared to P then to first order we have:

$$\Delta p^i(\underline{x}, t) = \frac{1}{c^2} \frac{\partial^2 p^i(\underline{x}, t)}{\partial t^2} \tag{1.1}$$

where Δ is the Laplace operator and $c = \dfrac{1}{\sqrt{\kappa\rho}}$ is the wave velocity, see [1] chapter 6.

---

* Dipartimento di Matematica e Fisica, Università di Camerino - 62032 Camerino (Italy).

† Istituto di Matematica e Statistica, Università di Ancona - 60100 Ancona (Italy).

‡ Dipartimento di Matematica "G. Castelnuovo", Università di Roma "La Sapienza" - 00185 Roma (Italy).

The obstacle D hit by the incoming wave, generates a scattered acoustic wave whose acoustic pressure is denoted by $p^s(\underline{x},t)$, $\underline{x} \in \mathbf{R}^3 \setminus \overline{D}$, $t > 0$. We suppose that the incoming acoustic wave is a time harmonic wave, that is:

$$p^i(\underline{x},t) = u^i(\underline{x})e^{-i\omega t}. \tag{1.2}$$

The scattered acoustic wave is also assumed to be time harmonic as in (1.2) and satisfies (1.1), we denote by $u^s(\underline{x})$, $\underline{x} \in \mathbf{R}^3 \setminus \overline{D}$ the space dependent part of $p^s(\underline{x},t)$. Let

$$u(\underline{x}) = u^i(\underline{x}) + u^s(\underline{x}) \;, \quad \underline{x} \in \mathbf{R}^3 \setminus \overline{D} \tag{1.3}$$

be the space dependent part of the total acoustic pressure outside the scatterer. From (1.1) it follows that u satisfies the Helmholtz equation:

$$\Delta u(\underline{x}) + k^2 u(\underline{x}) = 0 \;, \quad \underline{x} \in \mathbf{R}^3 \setminus \overline{D} \tag{1.4}$$

where $k = \frac{\omega}{c} > 0$ is the wave number. The scattered wave u^s depends on the shape and the acoustic properties of the obstacle D. We assume that the behaviour of the obstacle when hit by the incoming pressure field can be described by the following boundary condition (see [1] chapter 6):

$$u(\underline{x}) + \frac{\chi(\underline{x})}{ik}\frac{\partial u(\underline{x})}{\partial \underline{\hat{n}}(\underline{x})} = 0 \;, \quad \underline{x} \in \partial D \tag{1.5}$$

where $\underline{\hat{n}}(\underline{x})$ is the unit outward normal vector to ∂D at the point $\underline{x} \in \partial D$ and the complex function $\chi : \partial D \to \mathbf{C}$ is the boundary acoustic impedance that we simply call acoustic impedance. As limit cases of condition (1.5) we have the Dirichlet boundary condition, i.e. $\chi = 0$, that describes acoustically soft obstacles and the Neumann boundary condition, i.e. $\chi = \infty$, that decribes acoustically hard obstacles.

The scattered acoustic wave $u^s(\underline{x})$ has the asymptotic behaviour, for $\|\underline{x}\| \to \infty$, of an outgoing spherical wave. Hence u^s satisfies the Sommerfeld radiation condition:

$$\lim_{\|\underline{x}\|\to\infty} \|\underline{x}\| \left(\frac{\partial u^s(\underline{x})}{\partial \underline{\hat{x}}} - iku^s(\underline{x}) \right) = 0 \tag{1.6}$$

where $\underline{\hat{x}} = \dfrac{\underline{x}}{\|\underline{x}\|}$, $\underline{x} \neq 0$, see [2] page 188.

Let ∂D and χ be sufficiently regular, and let $Re(\chi) \leq 0$, where $Re(\cdot)$ denotes the real part of the complex number $\cdot$, then the scattered wave u^s is the unique solution of the boundary value problem (1.4), (1.5), (1.6), see [3] page 75 or [4] page 45.

Moreover every function u^s satisfying (1.4), (1.6) has the following expansion:

$$u^s(\underline{x}) = \frac{e^{ik\|\underline{x}\|}}{\|\underline{x}\|} \sum_{j=0}^{\infty} \frac{1}{\|\underline{x}\|^j} u_j(\underline{\hat{x}}). \tag{1.7}$$

This expansion holds for $||\underline{x}||$ sufficiently large and u_0 is known as acoustic far field pattern, see [3] page 72.

Let $B = \{\underline{x} \in \mathbf{R}^3 \mid ||\underline{x}|| < 1\}$ and ∂B be the boundary of B. We consider the following scattering problems:

Problem 1.1 (Acoustic direct problem) Let u^i be an incoming acoustic plane wave, that is:

$$u^i(\underline{x}) = e^{ik(\underline{x},\underline{\alpha})} \tag{1.8}$$

where $k > 0$ is the wave number and $\underline{\alpha}$, with $||\underline{\alpha}|| = 1$, is the wave propagation direction. Given the obstacle D, the acoustic impedance $\chi(\underline{x})$, $\underline{x} \in \partial D$ and the incoming acoustic field (1.8), determine the scattered acoustic field u^s solution of (1.4), (1.5), (1.6). In particular determine the corresponding far field pattern $u_0(\hat{\underline{x}}, k, \underline{\alpha})$, $\hat{\underline{x}} \in \partial B$.

Let λ_n, $n = 1, 2, \ldots$ be the eigenvalues of the Laplace operator in the domain D with the boundary condition (1.5).

Problem 1.2 (Acoustic inverse problem) Let $\Xi \subseteq \partial B$, $\Omega_1 \subset \mathbf{R}$, $\Omega_2 \subset \partial B$ be three given sets such that $(-\lambda_n)^{1/2} \notin \Omega_1$, $n = 1, 2, \ldots$. From a knowledge of the acoustic impedance χ and from a knowledge of $u_0(\hat{\underline{x}}, k, \underline{\alpha})$ for $\hat{\underline{x}} \in \Xi$, $k \in \Omega_1$, $\underline{\alpha} \in \Omega_2$, determine the boundary of the obstacle ∂D.

Problem 1.3 (Acoustic inverse problem) Let $\Xi \subseteq \partial B$, $\Omega_1 \subset \mathbf{R}$, $\Omega_2 \subseteq \partial B$ be three given sets such that $(-\lambda_n)^{1/2} \notin \Omega_1$, $n = 1, 2, \ldots$. From a knowledge of $u_0(\hat{\underline{x}}, k, \underline{\alpha})$ for $\hat{\underline{x}} \in \Xi$, $k \in \Omega_1$, $\underline{\alpha} \in \Omega_2$, determine the boundary of the obstacle ∂D and the acoustic impedance χ.

We note that $\lambda_n \in \mathbf{C}$, $n = 1, 2, \ldots$ and in Problem 1.2, 1.3 we have denoted by $(-\lambda_n)^{1/2}$ the principal branch of the square root of $-\lambda_n$ in $\mathbf{C}$. The condition $(-\lambda_n)^{1/2} \notin \Omega_1$, $n = 1, 2, \ldots$ is a non-resonance condition that is always satisfied for $Im(\chi) \neq 0$ and k real, where $Im(\cdot)$ denotes the imaginary part of the complex number $\cdot$.

Moreover Ω_2 is the set of the propagation directions of the incoming waves and Ξ is the set of the directions where the far field patterns are measured.

The electromagnetic scattering problem. We consider $\mathbf{R}^3 \backslash D$ filled by a homogeneous isotropic medium characterized by electric permittivity ϵ, magnetic permeability μ and zero electric conductivity. Moreover we assume ϵ, μ to be constants and the free charge density to be zero. The electric field $\underline{\mathcal{E}}^i(\underline{x}, t)$ and magnetic field $\underline{\mathcal{H}}^i(\underline{x}, t)$, $\underline{x} \in \mathbf{R}^3 \setminus \overline{D}$, $t > 0$ describing an incoming electromagnetic wave satisfy the Maxwell equations. When the incoming electromagnetic wave hits D, it generates a scattered electromagnetic field $\underline{\mathcal{E}}^s(\underline{x}, t)$, $\underline{\mathcal{H}}^s(\underline{x}, t)$, $\underline{x} \in \mathbf{R}^3 \setminus \overline{D}$, $t > 0$.

If the incoming electromagnetic wave is time harmonic, that is:

$$\underline{\mathcal{E}}^i(\underline{x},t) = \frac{1}{\sqrt{\epsilon}}\underline{E}^i(\underline{x})e^{-i\omega t}\ , \quad \underline{\mathcal{H}}^i(\underline{x},t) = \frac{1}{\sqrt{\mu}}\underline{H}^i(\underline{x})e^{-i\omega t} \tag{1.9}$$

we assume that the scattered electromagnetic field $\underline{\mathcal{E}}^s$, $\underline{\mathcal{H}}^s$ is also of the form (1.9) (and satisfy the Maxwell equations). Let $\underline{E}^s(\underline{x})$, $\underline{H}^s(\underline{x})$ be the space dependent part of $\underline{\mathcal{E}}^s(\underline{x},t)$, $\underline{\mathcal{H}}^s(\underline{x},t)$ respectively. We denote by $\underline{E}$, $\underline{H}$ the space dependent part of the total electromagnetic field outside the scatterer, that is:

$$\underline{E}(\underline{x}) = \underline{E}^i(\underline{x}) + \underline{E}^s(\underline{x})\ , \quad \underline{x} \in \mathbf{R}^3 \setminus \overline{D} \tag{1.10}$$

$$\underline{H}(\underline{x}) = \underline{H}^i(\underline{x}) + \underline{H}^s(\underline{x})\ , \quad \underline{x} \in \mathbf{R}^3 \setminus \overline{D}. \tag{1.11}$$

From the Maxwell equations it follows that the fields $\underline{E}$, $\underline{H}$ satisfy the time harmonic Maxwell equations, that is:

$$\mathbf{curl}\underline{E}(\underline{x}) - ik\underline{H}(\underline{x}) = \underline{0}\ , \quad \underline{x} \in \mathbf{R}^3 \setminus \overline{D} \tag{1.12}$$

$$\mathbf{curl}\underline{H}(\underline{x}) + ik\underline{E}(\underline{x}) = \underline{0}\ , \quad \underline{x} \in \mathbf{R}^3 \setminus \overline{D} \tag{1.13}$$

where $k = \omega\sqrt{\mu\epsilon} > 0$ is the wave number. Moreover by straightforward calculations we have that the vector field $\underline{E}$ satisfies the vector Helmholtz equation and is a divergence free field, that is:

$$\Delta\underline{E}(\underline{x}) + k^2\underline{E}(\underline{x}) = \underline{0}\ , \quad \underline{x} \in \mathbf{R}^3 \setminus \overline{D} \tag{1.14}$$

$$\mathbf{div}\underline{E}(\underline{x}) = 0\ , \quad \underline{x} \in \mathbf{R}^3 \setminus \overline{D} \tag{1.15}$$

where Δ is the "vector" Laplace operator and "**div**" denotes the divergence operator. Also the field $\underline{H}$ satisfies (1.14), (1.15) , see [4] page 147 for a detailed discussion.

The scattered field $\underline{E}^s$ depends on the shape and the electric properties of the scatterer. We assume that the electric properties of the scatterer are expressed by the following boundary condition:

$$[\underline{\hat{n}}(\underline{x}), \underline{E}(\underline{x})] + \frac{\psi(\underline{x})}{ik}[\underline{\hat{n}}(\underline{x}), [\underline{\hat{n}}(\underline{x}), \mathbf{curl}\underline{E}(\underline{x})]] = \underline{0}\ , \quad \underline{x} \in \partial D \tag{1.16}$$

where $\psi : \partial D \to \mathbf{C}$ is the electric boundary impedance of the obstacle, see [5] page 532. As limit cases of condition (1.16) we have a perfectly conducting obstacle for $\psi = 0$ and a perfectly insulating obstacle for $\psi = \infty$.

The scattered electromagnetic field has the asymptotic behaviour at infinity of a spherical electromagnetic wave outgoing from the origin of the coordinate system. Thus we impose the following condition:

$$\lim_{\|\underline{x}\|\to\infty} \|\underline{x}\| \left(\frac{1}{ik}[\mathbf{curl}\underline{E}^s(\underline{x}), \underline{\hat{x}}] - \underline{E}^s(\underline{x})\right) = \underline{0} \tag{1.17}$$

known as Silver-Müller radiation condition, see [4] page 153.

Let ∂D and ψ be sufficiently regular and let $Re(\psi) \leq 0$, then the scattered field $\underline{E}^s$ is the unique solution of the boundary value problem (1.14), (1.15), (1.16), (1.17), see [3] page 121.

Moreover the scattered electric field $\underline{E}^s$, solution of (1.14), (1.15), (1.17) has the following expansion, see [3] page 116:

$$\underline{E}^s(\underline{x}) = \frac{e^{ik\|\underline{x}\|}}{\|\underline{x}\|} \sum_{j=0}^{\infty} \frac{1}{\|\underline{x}\|^j} \underline{E}_j(\hat{\underline{x}}). \tag{1.18}$$

This expansion holds for $\|\underline{x}\|$ sufficiently large and $\underline{E}_0$ is known as the electric far field pattern.

Problem 1.4 (Electric direct problem) Given a linearly polarized electric plane wave:

$$\underline{E}^i(\underline{x}) = \underline{w}\, e^{ik(\underline{x},\underline{\alpha})} \tag{1.19}$$

where $k > 0$ is the wave number, $\underline{\alpha}$, with $\|\underline{\alpha}\| = 1$, is the wave propagation direction and $\underline{w}$, with $(\underline{w}, \underline{\alpha}) = 0$, is the polarization vector. Given an obstacle D and the electric boundary impedance $\psi(\underline{x})$, $\underline{x} \in \partial D$, determine the scattered electric field, in particular determine the corresponding electric far field pattern $\underline{E}_0(\hat{\underline{x}}, k, \underline{\alpha}, \underline{w})$, $\hat{\underline{x}} \in \partial B$.

Let $\mu_n, n = 1, 2, \ldots$ be the eigenvalues of the vector Laplace operator restricted to the divergence free vector fields in the interior of D with the boundary condition (1.16).

Problem 1.5 (Electric inverse problem) Let $\Xi \subseteq \partial B$, $\Omega_1 \subset \mathbf{R}$, $\Omega_2 \subset \partial B \times \mathbf{R}^3$, be three given sets such that $(-\mu_n)^{1/2} \notin \Omega_1$, $n = 1, 2, \ldots$. From a knowledge of the electric boundary impedance ψ of the obstacle and from a knowledge of $\underline{E}_0(\hat{\underline{x}}, k, \underline{\alpha}, \underline{w}_{\underline{\alpha}})$, for $\hat{\underline{x}} \in \Xi$, $k \in \Omega_1$, $(\underline{\alpha}, \underline{w}_{\underline{\alpha}}) \in \Omega_2$, determine the shape ∂D of the obstacle D.

Problem 1.6 (Electric inverse problem) Let $\Xi \subseteq \partial B$, $\Omega_1 \subset \mathbf{R}$, $\Omega_2 \subset \partial B \times \mathbf{R}^3$, be three given sets such that $(-\mu_n)^{1/2} \notin \Omega_3$, $n = 1, 2, \ldots$. From a knowledge of $\underline{E}_0(\hat{\underline{x}}, k, \underline{\alpha}, \underline{w}_{\underline{\alpha}})$, for $\hat{\underline{x}} \in \Xi$, $k \in \Omega_1$, $(\underline{\alpha}, \underline{w}_{\underline{\alpha}}) \in \Omega_2$, determine the boundary ∂D of the obstacle and the electric boundary impedance ψ.

Problems similar to Problems 1.5, 1.6 can be considered for the magnetic field.

The direct problems (i.e. Problem 1.1 and Problem 1.4) are well-posed, that is given the shape ∂D and the acoustic impedance χ (or ψ in the electromagnetic case) of the obstacle D, the scattered field u^s ($\underline{E}^s$) and the corresponding far field pattern exists and are unique. Moreover u^s ($\underline{E}^s$) or the far field u_0 ($\underline{E}_0$) depends continuously on the data $\partial D, \chi$.

The inverse problems (i.e. Problems 1.2, 1.3, 1.5, 1.6) are ill-posed, see [4] page 104 for a detailed discussion on ill-posedness of the inverse problem.

Moreover the direct problems are linear while the inverse problems are nonlinear.

In this paper we present two different methods to solve the inverse problems presented previously: the iterative solution of the corresponding direct problem that we refer as the *Direct problem approach* and the so called *Dual space method.* The last method avoids the solution of the corresponding direct problem that is computationally expensive. So we prefer to solve inverse problems by the *Dual space method.* However the use of multiprocessor computers and a highly parallelizable algorithm to solve the direct problems can make convenient the solution of inverse problems by iterative solution of the corresponding direct problem.

In this paper we are mainly concerned with k in the resonance region, that is:

$$kL \approx 1 \tag{1.20}$$

where L is a characteristic length of the obstacle D. In these circumstances perturbation techniques, that is low frequency approximations ($kL \ll 1$) or high frequency approximations ($kL \gg 1$), are not available.

In Section 2 we present two numerical methods to solve the direct problem. Such methods are necessary to obtain synthetic data for the inverse problem; moreover the inverse problem can be solved by iterative solution of the direct problem. Thus a parallel algorithm for the direct problem can be used to solve efficiently the inverse problem. In Section 3 we consider two approaches to solve the inverse problem leading to unconstrained optimization problems. In Section 4 we explain the use of optimization in the numerical solution of the inverse problem and we show some numerical results. In Section 5 some conclusions are drawn and areas of future work are suggested.

2. The direct problem. There are many numerical methods to solve Problems 1.1, 1.4, for example it is possible to use finite differences or finite elements methods in a bounded domain containing the obstacle employing absorbing boundary conditions on the artificial boundaries, see [6]. A method particularly suited for scattering problems is the so called T-matrix method, see [7]. We explain this method in the acoustic case, that is for Problem 1.1, but in a similar manner it can be used to compute the solution of the electric direct problem, that is Problem 1.4, see [8]. Moreover the T-matrix method can be generalized to compute, for example, the scattered field from non connected obstacles, see [9], or from random media, see [10].

The T-matrix method consists of constructing an operator $T : C^2(\mathbf{R}^3 \setminus \overline{D}) \to C^2(\mathbf{R}^3 \setminus \overline{D})$ that depends only on the obstacle D and the boundary condition on ∂D and that gives the scattered field u^s in terms of the incoming field u^i, that is:

$$(Tu^i)(\underline{x}) = u^s(\underline{x}) \ , \quad \underline{x} \in \mathbf{R}^3 \setminus \overline{D}. \tag{2.1}$$

In the numerical treatment we consider a truncated expansion of the incoming field and of the scattered field, that is:

$$u^i(\underline{x}) = \sum_{\sigma=0,1} \sum_{l=\sigma}^{L_{max}} \sum_{m=\sigma}^{l} a_{\sigma,m,l}\ Re\varphi_{\sigma,m,l}(\underline{x}), \quad \underline{x} \in \mathbf{R}^3 \tag{2.2}$$

$$u^s(\underline{x}) = \sum_{\sigma=0,1} \sum_{l=\sigma}^{L_{max}} \sum_{m=\sigma}^{l} f_{\sigma,m,l}\ \varphi_{\sigma,m,l}(\underline{x}), \quad \underline{x} \in \mathbf{R}^3 \setminus \overline{D}. \tag{2.3}$$

where L_{max} is the truncation parameter and (ρ, θ, ϕ) denotes the usual spherical coordinates of $\mathbf{R}^3$. The bases:

$$\begin{array}{llll} \{\varphi_{\sigma,m,l} \,|\, \sigma & = 0,1, & l = \sigma, \sigma+1, \ldots, & m = \sigma, \ldots, l\} \\ \{Re\,\varphi_{\sigma,m,l} \,|\, \sigma & = 0,1, & l = \sigma, \sigma+1, \ldots, & m = \sigma, \ldots, l\} \end{array} \tag{2.4}$$

are respectively the basis of the wave functions and the basis of the regular wave functions, i.e. $\varphi_{\sigma,m,l}(\underline{x}) = h_l^{(1)}(k\rho) Y_{\sigma,m,l}(\theta,\phi)$ and $Re\,\varphi_{\sigma,m,l}(\underline{x}) = j_l(k\rho) Y_{\sigma,m,l}(\theta,\phi)$, where $h_l^{(1)}$ are the spherical Hankel functions of first kind, j_l are the spherical Bessel functions and $Y_{\sigma,m,l}$ are the spherical harmonics, that is:

$$Y_{\sigma,m,l}(\theta,\phi) = \gamma_{m,l}\ P_l^m(\cos\theta)\ \cos\big(m\phi - \sigma\frac{\pi}{2}\big) \tag{2.5}$$

where P_l^m are the Legendre functions, $\gamma_{m,l} = \sqrt{\frac{\varepsilon_m (2l+1)(l-m)!}{4\pi(l+m)!}}$ and $\varepsilon_m = 2$ if $m = 0$, $\varepsilon_m = 1$ if $m \neq 0$, see [11] page 1462 for more details.

Taking advantage of these expansions the operator T becomes a matrix that gives the coefficients of expansion (2.3) in terms of the ones of (2.2).

The T-matrix elements are now solutions of a linear system, see [7] for more details. Moreover if the incoming acoustic field is a plane wave the coefficients of the expansion (2.2) are given by an explicit formula, see [11] page 1466.

The advantage of the T-matrix method is that the computation of the T-matrix, which is the computationally most expensive operation in the solution of Problem 1.1, must be performed only once even if we have to calculate the scattered fields generated by several incoming waves with the same wave number k.

In [12] Milder has developed a new method to compute the acoustic wave scattered from an open unbounded obstacle. This suggests the generalization of the method proposed by Milder to bounded simply connected obstacles [13].

To fix the ideas we introduce the method for an acoustically soft obstacle, i.e. we assume the boundary condition (1.5) with $\chi = 0$. The main step of this method is the construction of a nonlocal operator $\hat{\mathcal{N}}$ such that:

$$(\hat{\mathcal{N}} u^s)(\underline{x}) = \frac{\partial u^s(\underline{x})}{\partial \underline{\hat{n}}(\underline{x})}, \quad \underline{x} \in \partial D. \tag{2.6}$$

The Helmholtz formula gives:

$$\int_{\partial D}\left(u^s(\underline{y})\frac{\partial\Phi(\underline{x},\underline{y})}{\partial\underline{\hat{n}}(\underline{y})}-\frac{\partial u^s(\underline{y})}{\partial\underline{\hat{n}}(\underline{y})}\Phi(\underline{x},\underline{y})\right)ds(\underline{y}) = \begin{cases} u^s(\underline{x}), & \underline{x}\in\mathbf{R}^3\setminus\overline{D} \\ 0, & \underline{x}\in D \end{cases} \tag{2.7}$$

where

$$\Phi(\underline{x},\underline{y})=\frac{e^{ik\|\underline{x}-\underline{y}\|}}{4\pi\,\|\underline{x}-\underline{y}\|}\ ,\ \underline{x},\underline{y}\in\mathbf{R}^3\ ,\ \underline{x}\neq\underline{y} \tag{2.8}$$

is the free space Green's function of the Helmholtz operator with the Sommerfeld radiation condition at infinity and ds is the surface measure on ∂D. Using the boundary condition (1.5) with $\chi=0$, we can compute the acoustic scattered wave $u^s(\underline{x})$ for every $\underline{x}\in\mathbf{R}^3\setminus\overline{D}$, that is:

$$u^s(\underline{x})=-\int_{\partial D}\left(u^i(\underline{y})\frac{\partial\Phi(\underline{x},\underline{y})}{\partial\underline{\hat{n}}(\underline{y})}-(\hat{\mathcal{N}}u^i)(\underline{y})\Phi(\underline{x},\underline{y})\right)ds(\underline{y}),\quad \underline{x}\in\mathbf{R}^3\setminus\overline{D}. \tag{2.9}$$

To fix the ideas let us consider the usual spherical coordinates in $\mathbf{R}^3$, (ρ,θ,ϕ), we suppose D to be a star-like domain with respect to the origin and let $\rho=f(\theta,\phi)$, $0\le\theta\le\pi$, $0\le\phi<2\pi$ be a parametrization of the boundary ∂D of D. Then the operator $\hat{\mathcal{N}}$ depends on f and we can consider a formal expansion of the operator $\hat{\mathcal{N}}$ in "powers" of f, that is:

$$\hat{\mathcal{N}}=\sum_{i=0}^{\infty}\hat{\mathcal{N}}_i. \tag{2.10}$$

Roughly speaking, the $\hat{\mathcal{N}}_i$ are defined by integrals involving the solution of the following integral equation:

$$\int_{\partial B}\Phi(\underline{x},\underline{\hat{y}})\,v(\underline{\hat{y}})d\sigma(\underline{\hat{y}})\ =\ -u^i(\underline{x}),\ \underline{x}\in\partial D\,, \tag{2.11}$$

where $d\sigma$ is the surface measure on ∂B. The numerical solution of (2.11) is obtained perturbatively. That is, let $\underline{\hat{x}}$, $\underline{\hat{y}}\in\partial B$, the integral equation (2.11) is solved expanding $\Phi(f(\theta,\phi)\underline{\hat{x}},\underline{\hat{y}})$ and $v(\underline{\hat{y}})$ in powers of $f(\theta,\phi)-1$. So that the computation of the operators $\hat{\mathcal{N}}_i$ involves only integrals independent one from the other, and no linear systems are involved.
A numerical approximation of u^s, can be obtained by substituting (2.10) in (2.9) and truncating the series (2.10). The computation of the series resulting from substituting (2.10) in (2.9) corresponds to the computation of integrals whose integrands are independent one from the other so that this method is very well suited for parallel computation.

On the contrary the T-matrix method, and the classical numerical methods for boundary value problems, i.e. finite differences, finite elements, imply the solution of linear systems, this makes the T-matrix method and the other methods not well suited for massive parallelism.

It is very important to have an efficient algorithm to solve a direct scattering problem when solving the inverse scattering problem with the *Direct problem approach*.

3. The inverse problem as an optimization problem. The inverse Problems 1.2, 1.3, 1.5, 1.6 can be treated in a similar way, to fix the ideas we consider Problem 1.2. We present two approaches to obtain a numerical approximation of the solution of Problem 1.2.

The first one is the *Direct problem approach*. Given an incoming plane wave u^i of the form (1.8) with incoming direction $\underline{\alpha}$ and wave number k, we define the operator $F^{\underline{\alpha},k} : \mathcal{U} \to L^2(\partial B)$, where $\mathcal{U}$ is the space of all closed smooth surfaces, as the following map:

$$F^{\underline{\alpha},k} : \partial D \mapsto u_0(\cdot, k, \underline{\alpha}) \tag{3.1}$$

so that the evaluation of the operator $F^{\underline{\alpha},k}$ requires the solution of Problem 1.1.

Under appropriate hypotheses the map $F^{\underline{\alpha},k}$ is Fréchet differentiable with respect to ∂D (see [4] page 120), this suggests an algorithm to solve the inverse scattering problem. That is we can find an approximation Λ^* of ∂D as the solution of the following optimization problem:

$$\inf_{\Lambda \in \mathcal{U}} \left\| F^{\underline{\alpha},k}(\Lambda) - u_0(\cdot, k, \underline{\alpha}) \right\|^2_{L^2(\Xi)} . \tag{3.2}$$

We note that in the numerical treatment of problem (3.2), using a gradient method we have to use a regularization technique such as the Tikhonov one, see [14]. In fact the computation of the descent direction is an ill-posed problem, see [4] chapter 5.

The knowledge of an infinite number of far field patterns generated by the obstacle when hit by different incoming plane waves guarantees the uniqueness of the solution of the inverse problem, see [4] page 104. Thus we define a new optimization problem:

$$\inf_{\Lambda \in \mathcal{U}} \sum_{k \in \Omega_1} \sum_{\underline{\alpha} \in \Omega_2} \left\| F^{\underline{\alpha},k}(\Lambda) - u_0(\cdot, k, \underline{\alpha}) \right\|^2_{L^2(\Xi)} . \tag{3.3}$$

The second method we present has been introduced by Colton and Monk [15], [16] and avoids the solution of Problem 1.1 and it is known as *Dual space method*. We present this method for Problem 1.2 , but it can be generalized to solve Problem 1.3, 1.5, 1.6 too. See [17], [18], [19], [20], [21], [22] where the work of Colton and Monk has been generalized to solve the inverse problems presented in Section 1.

Let $k \in \Omega_1$ and let v^k be the unique solution of the boundary value problem:

$$\Delta v^k(\underline{y}) + k^2 v^k(\underline{y}) = 0 \ , \ \underline{y} \in D \tag{3.4}$$

$$v^k(\underline{y}) - \frac{\overline{\chi}(\underline{y})}{ik} \frac{\partial v^k(\underline{y})}{\partial \underline{\hat{n}}(\underline{y})} = \frac{4\pi}{k} \left(\overline{\Phi}(\underline{\xi}, \underline{y}) - \frac{\overline{\chi}(\underline{y})}{ik} \frac{\partial \overline{\Phi}(\underline{\xi}, \underline{y})}{\partial \underline{\hat{n}}(\underline{y})} \right) \Bigg|_{\underline{\xi}=\underline{0}} , \ \underline{y} \in \partial D \tag{3.5}$$

where the overbar means complex conjugate.

Definition 3.1 Let $d\sigma$ be the surface measure on ∂B. The domain D is said to be a Herglotz domain if the unique solution of problem (3.4), (3.5) can be represented as a Herglotz wave function, that is:

$$v^k(\underline{y}) = \int_{\partial B} g^k(\underline{\hat{x}}) e^{ik(\underline{\hat{x}}, \underline{y})} d\sigma(\underline{\hat{x}}) \ , \ \underline{y} \in \mathbf{R}^3 \tag{3.6}$$

for a suitable choice of $g^k \in L^2(\partial B)$. The function g^k is said to be the Herglotz kernel associated to D.

Let D be a Herglotz domain and g^k the corresponding kernel then it is easy to see that:

$$\int_{\partial B} u_0(\underline{\hat{x}}, k, \underline{\alpha}) \overline{g}^k(\underline{\hat{x}}) d\sigma(\underline{\hat{x}}) = \frac{1}{k} \ , \ \forall \underline{\alpha} \in \Omega_2. \tag{3.7}$$

For the Herglotz domains D Problem 1.2 can be solved now in three steps:

i) for any $k \in \Omega_1$ and from the knowledge of the far field patterns $u_0(\underline{\hat{x}}, k, \underline{\alpha})$, $\underline{\hat{x}} \in \Xi$, $\underline{\alpha} \in \Omega_2$, determine an approximation of g^k using (3.7);

ii) from g^k determine v^k, using (3.6);

iii) from v^k determine ∂D, using (3.5), that is we search an approximation of ∂D as the solution of the optimization problem:

$$\begin{aligned} \inf_{\Lambda \in \mathcal{U}} \sum_{k \in \Omega_1} \Bigg\| \Bigg\{ & v^k(\cdot) - \frac{\overline{\chi}(\cdot)}{ik} \frac{\partial v^k(\cdot)}{\partial \underline{\hat{n}}(\cdot)} \\ & - \frac{4\pi}{k} \left(\overline{\Phi}(\underline{x}, \cdot) - \frac{\overline{\chi}(\cdot)}{ik} \frac{\partial \overline{\Phi}(\underline{x}, \cdot)}{\partial \underline{\hat{n}}(\cdot)} \right) \Bigg|_{\underline{x}=\underline{0}} \Bigg\} \Bigg\|^2_{L^2(\Lambda)} . \end{aligned} \tag{3.8}$$

We note that the ill-posedness of the inverse problem is translated into the ill-posedness of (3.7) when we look at (3.7) as an equation defining g^k.

4. Some numerical experiments. In the numerical experiments described here we assume D to be a star-like domain with respect to the origin so that ∂D has a global parametrization in spherical coordinates (ρ, θ, ϕ), i.e. $\partial D = \{(\rho, \theta, \phi) \in \mathbf{R}^3 \,|\, \rho = f(\theta, \phi), 0 \le \theta \le \pi, 0 \le \phi < 2\pi\}$. We search for an approximation $\Lambda = \Lambda(\underline{c})$ of ∂D as a truncated series of spherical harmonics, that is:

$$\Lambda(\underline{c}) = \Big\{(\rho,\theta,\phi) \in \mathbf{R}^3 \mid \rho = \tilde{f}(\theta,\phi,\underline{c}), \tag{4.1}$$
$$\tilde{f}(\theta,\phi,\underline{c}) = \textstyle\sum_{\sigma=0,1}\sum_{l=\sigma}^{L_\rho}\sum_{m=\sigma}^{l} c_{\sigma,m,l} Y_{\sigma,m,l}\,(\theta,\phi), 0 \le \theta \le \pi, 0 \le \phi < 2\pi\Big\}$$

where $\underline{c} = (c_{\sigma,m,l} \mid \sigma = 0,1,\ l = \sigma,\ldots,L_\rho,\ m = \sigma,\ldots,l)$ and L_ρ is a truncation parameter.

In the *Direct problem approach* we can approximate the optimization problem (3.3) with the following finite dimensional optimization problem:

$$\inf_{\underline{c} \in \mathbf{R}^{(L_\rho+1)^2}} \sum_{k\in\Omega_1} \sum_{\underline{\alpha}\in\Omega_2} \left\| F^{\underline{\alpha},k}(\Lambda(\underline{c})) - u_0(\cdot,k,\underline{\alpha}) \right\|^2_{L^2(\Xi)} \,. \tag{4.2}$$

Let $\underline{c}^*$ be a solution of (4.2), we take as an approximation of ∂D the surface $\Lambda^* = \Lambda(\underline{c}^*)$ according to (4.1).

To apply efficiently the *Dual space method* it is convenient to approximate the far field pattern with a truncated series of spherical harmonics, that is:

$$u_0(\hat{\underline{x}},k,\underline{\alpha}) \approx \sum_{\sigma=0,1} \sum_{l=\sigma}^{L_M} \sum_{m=\sigma}^{l} f^{k,\underline{\alpha}}_{\sigma,m,l} Y_{\sigma,m,l}(\hat{\underline{x}}),\ \ \hat{\underline{x}} \in \partial B,\ k \in \Omega_1,\ \underline{\alpha} \in \Omega_2 \tag{4.3}$$

and for $k \in \Omega_1$, $\underline{\alpha} \in \Omega_2$ we obtain such an approximation solving the following linear system:

$$\sum_{\sigma=0,1} \sum_{l=\sigma}^{L_M} \sum_{m=\sigma}^{l} f^{k,\underline{\alpha}}_{\sigma,m,l} Y_{\sigma,m,l}(\hat{\underline{x}}) = u_0(\hat{\underline{x}},k,\underline{\alpha})\ ,\ \hat{\underline{x}} \in \Xi \tag{4.4}$$

where we must choose the parameter L_M such that $(L_M+1)^2 \le \mathrm{card}(\Xi)$. The notation $\mathrm{card}(\cdot)$ denotes the cardinality of the set $\cdot$. The linear system (4.4) is solved in the least squares sense.

The *Dual space method* to solve the inverse problem is reduced to the following three steps algorithm:

Step $\imath$ is the solution of a linear system, in fact let us suppose that we want to compute an approximation of the Herglotz kernel g^k in the following form:

$$g^k(\hat{\underline{x}}) \approx \sum_{\sigma=0,1} \sum_{l=\sigma}^{L_g} \sum_{m=\sigma}^{l} g^k_{\sigma,m,l} Y_{\sigma,m,l}(\hat{\underline{x}}) \tag{4.5}$$

then from (3.7), (4.3) and from the orthogonality properties of the spherical harmonics, for $k \in \Omega_1$ we obtain:

$$\sum_{\sigma=0,1} \sum_{l=\sigma}^{L_g} \sum_{m=\sigma}^{l} f^{k,\underline{\alpha}}_{\sigma,m,l} \overline{g}^k{}_{\sigma,m,l} = \frac{1}{k}\ ,\ \underline{\alpha} \in \Omega_2 \tag{4.6}$$

where we assume $L_g \leq L_M$ and $(L_g+1)^2 \leq \text{card}(\Omega_2)$. Let $\underline{g}^k = (g^k_{\sigma,m,l} \mid \sigma = 0,1, l = \sigma,\ldots,L_g, m = \sigma,\ldots,l)$, $k \in \Omega_1$ be the solution in the least squares sense of (4.6).

Step *ii* is from formulas (3.6), (4.5), see [11] page 1467, we have the following approximation for the corresponding Herglotz wave function v^k:

$$(4.7)\; v^k(\underline{y}) \approx 4\pi \sum_{\sigma=0,1} \sum_{l=\sigma}^{L_g} \sum_{m=\sigma}^{l} i^l g^k_{\sigma,m,l} Y_{\sigma,m,l}(\underline{\hat{y}}) j_l(k\,\|y\|)\ ,\ \underline{y} \in \mathbf{R}^3\ ,\ k \in \Omega_1.$$

Step *iii* from (4.1) we can reformulate (3.8) as a finite dimensional optimization problem:

$$(4.8)\quad \begin{aligned} \inf_{\underline{c} \in \mathbf{R}^{(L_\rho+1)^2}} \sum_{k\in\Omega_1} \int_{\partial B} \Bigg| & v^k(\underline{y}(\underline{\hat{x}},\underline{c})) - \frac{\overline{\chi}(\underline{y}(\underline{\hat{x}},\underline{c}))}{ik} \frac{\partial v^k(\underline{y}(\underline{\hat{x}},\underline{c}))}{\partial \underline{\hat{n}}(\underline{y}(\underline{\hat{x}},\underline{c}))} - \\ & -\frac{4\pi}{k}\left(\overline{\Phi}(\underline{\xi},\underline{y}(\underline{\hat{x}},\underline{c})) - \frac{\overline{\chi}(\underline{y}(\underline{\hat{x}},\underline{c}))}{ik} \frac{\partial \overline{\Phi}(\underline{\xi},\underline{y}(\underline{\hat{x}},\underline{c}))}{\partial \underline{\hat{n}}(\underline{y}(\underline{\hat{x}},\underline{c}))}\right)\Bigg|_{\underline{\xi}=\underline{0}}\Bigg|^2 d\sigma(\underline{\hat{x}}) \end{aligned}$$

where $\underline{y}(\underline{\hat{x}},\underline{c}) = \tilde{f}(\underline{\hat{x}},\underline{c})\underline{\hat{x}}$.

We note that every a-priori information on the shape of the obstacle can be used to add penalization terms to the optimization problems (4.2), (4.8).

We note that the *Dual space method* allows us to handle efficiently also more general a-priori information on the obstacle shape. For example we often have a-priori information on the symmetries of ∂D at some special points, such as:

$$(4.9)\qquad \frac{\partial f}{\partial \theta}(\underline{\hat{y}}) = \frac{\partial f}{\partial \phi}(\underline{\hat{y}}) = 0\ ,\ \underline{\hat{y}} \in \Omega_4$$

where $\Omega_4 \subset \partial B$ is a given set. For example $\Omega_4 = \{(0,0,1)\,(\text{north pole}), (0,0,-1)\,(\text{south pole})\}$. Equation (3.5) defines a relation among f, $\frac{\partial f}{\partial \theta}$ and $\frac{\partial f}{\partial \phi}$. For $\underline{\hat{y}} \in \Omega_4$, using (4.9), equation (3.5) becomes a nonlinear equation for $f(\underline{\hat{y}})$, so that we can solve it accurately. Let $\tilde{f}^*_{\underline{\hat{y}}}$, $\underline{\hat{y}} \in \Omega_4$ be the solution of (3.5). This new information on the shape of the obstacle can be used adding to the objective function of problem (4.8), the following penalization term:

$$(4.10)\qquad \sum_{\underline{\hat{y}}\in\Omega_4} W_{\underline{\hat{y}}} \left|\tilde{f}(\underline{\hat{y}},\underline{c}) - \tilde{f}^*_{\underline{\hat{y}}}\right|^2$$

where $W_{\underline{\hat{y}}}$ are positive weights.

The computational cost of the *Dual space method* is equal to the computational cost of the solution of the linear systems (4.4), (4.6) plus the computational cost of problem (4.8). We analyze now the computational cost of the two algorithms presented that is the *Dual space method* and the *Direct problem approach* considering only the computational cost of the function evaluation of the corresponding optimization problems. We can say that the computational cost of the *Direct problem approach* is higher than the one of the *Dual space method* when the direct scattering problem is solved by classical numerical methods, such as the T-matrix method, a finite differences methods, that both require the solution of several large linear systems at every evaluation of the objective function of problem (4.2). However with a fully parallelizable algorithm to solve the direct problem (see Section 2), and using a multiprocessor computer the cost of these two algorithms can become comparable and eventually the *Direct problem approach* may become convenient with respect the *Dual space method.* The interesting feature of the *Direct problem approach* is its generality, i.e. every inverse problem can be treated with this method

The *Dual space method* is used in [18], [20] to solve Problems 1.2, 1.5 and these inverse problems are always reformulated as global optimization problems, i.e. the independent variables of the optimization problem are the coefficients $\underline{c}$ introduced in (4.1). In [19] Problem 1.3 is reformulated as several pointwise optimization problems, i.e. the variables of each optimization problem are f and χ for a given direction $\hat{\underline{x}}$. The resulting algorithm is very fast. We note that the *Direct problem approach* does not allow such a reformulation of the inverse problem. In [22] Problem 1.6 is reformulated as a two steps optimization procedure: in the first one, via a global optimization problem, is reconstructed the obstacle shape, in the second one, from the knowledge of the obstacle shape, the electric boundary impedance of the scatterer is reconstructed as the solution of a pointwise optimization problem.

In [17], [18], [19], [20], [21], [22] there are many numerical experiments on the inverse scattering problems considered in Section 1. The data of these problems, i.e. the far field patterns, are generated by the T-matrix method. They can be more accurate than the data obtainable by instrumental measurements of the far field patterns generated by the obstacle, therefore in the reconstruction procedure these data are perturbed by random errors.

We give here one example for each inverse problem considered in Section 1.

Example 4.1 We consider Problem 1.2 for the so called Pseudo Apollo:

$$\partial D = \left\{ (\rho, \theta, \phi) \in \mathbf{R}^3 \;\middle|\; \rho = \frac{3}{5}\sqrt{\frac{17}{4} + 2\cos 3\theta}\ ,\ \ 0 \leq \theta \leq \pi,\ 0 \leq \phi < 2\pi \right\} \tag{4.11}$$

with the Dirichlet boundary condition, i.e. condition (1.5) with $\chi = 0$.

The far field patterns is obtained from formula (2.3) using a truncation parameter $L_{max} = L_M = 8$ as explained in Section 2. Moreover the inverse problem is solved using perturbed far field patterns $\tilde{u}_0$, that is:

$$\tilde{u}_0 = u_0 + \epsilon\zeta|u_0| \tag{4.12}$$

where $\epsilon > 0$ and ζ is a random number uniformly distributed in $[-1, 1]$.

The sets Ξ, Ω_2 are taken as uniform meshes on ∂B in the variables θ, ϕ and $\Omega_1 = \{3\}$.

FIG. 1. *Original* *Reconstructed*

$\epsilon = 0.05, \ E^f_{L^2} \approx 0.114$

The reconstrunction is performed solving the unconstrained optimization problem (4.8) with $L_\rho = 4$ using the a-priori information on the axial-symmetry of the obstacle. This leads to an optimization problem with 5 independent variables, i.e. the coefficients $c_{0,0,l}$ $l = 0, 1, 2, 3, 4$ given in formula (4.1). The sphere of radius 1 is always used as the initial guess of the optimization procedure. Moreover to solve problem (4.8) the routine SIGMA is used. The routine SIGMA is based on a stochastic global optimization algorithm, see [23].

The following parameters are used in the procedure to solve the inverse problem: $L_{max} = L_M = 8$, $L_g = 8$. The results obtained are shown in Fig. 1 and are taken from [18].

Example 4.2 We consider Problem 1.3 for a cylinder-like obstacle:

$$\partial D = \left\{ (\rho, \theta, \phi) \in \mathbf{R}^3 \,\middle|\, \rho = \left[\sqrt[10]{\left(\tfrac{2}{3}\sin\theta\right)^{10} + \cos^{10}\theta} \right]^{-1}, \quad 0 \le \theta \le \pi, \ 0 \le \phi < 2\pi \right\} \tag{4.13}$$

and a piecewise constant impedance:

$$\chi(\theta) = \begin{cases} -1, & 0 \le \theta < \frac{\pi}{2} \\ -2, & \frac{\pi}{2} \le \theta \le \pi. \end{cases} \tag{4.14}$$

The far field patterns are computed using the T-matrix method with a truncation parameter $L_{max} = L_M = 8$.

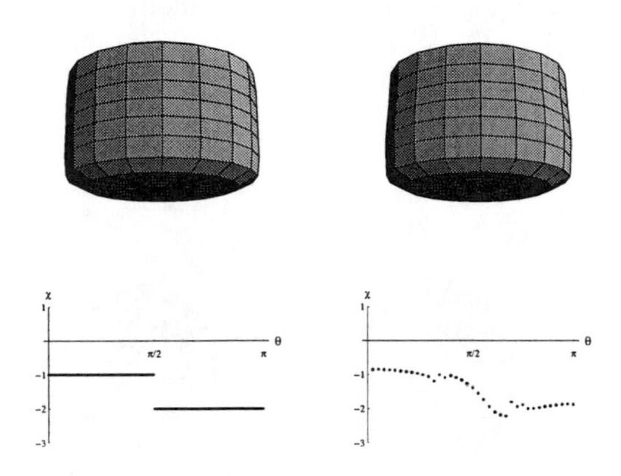

FIG. 2. *Original* *Reconstructed*

$E^f_{L^2} \approx 0.022, \;\; E^{\chi}_{L^2} \approx 0.101$

The sets Ξ, Ω_2 are taken as uniform meshes on ∂B in the variables θ, ϕ and $\Omega_1 = \{2, 3\}$.

The reconstruction is done using the axial-symmetry of the obstacle as a-priori information. The inverse problem is reformulated as several local unconstrained optimization problems. In particular the i-th problem has f_i, χ_i, $i = 0, \ldots, N$ as independent variables to be determined, where $f_i = f(\theta_i)$, $\chi_i = \chi(\theta_i)$ and $\theta_i = i\frac{\pi}{N}$, $N = 36$. Moreover to solve this optimization problem we use the routine DUNLSJ based on the Levenberg-Marquardt algorithm [24]. The initial guesses used are $f(\theta) = 1$ and $\chi(\theta) = 1$, $0 \le \theta \le \pi$. The results obtained are shown in Fig. 2 and are taken from [19].

Example 4.3 We consider Problem 1.5 for the ellipsoid:

$$\partial D = \left\{ (\rho, \theta, \phi) \in \mathbf{R}^3 \;\middle|\; \rho = \left[\sqrt{\left(1 - \tfrac{7}{16}\cos^2\phi\right)\sin^2\theta + \left(\tfrac{2}{3}\cos\theta\right)^2} \right]^{-1}, \; 0 \le \theta \le \pi, \;\; 0 \le \phi < 2\pi \right\} \tag{4.15}$$

with the electric boundary impedance $\psi(\underline{x}) = \frac{-1+i}{2}$, $\underline{x} \in \partial D$.

The inverse problem is solved, as in Example 4.1, with noisy data, that is $\underline{E}_0 = (\underline{E}_0, \hat{\underline{\theta}})\hat{\underline{\theta}} + (\underline{E}_0, \hat{\underline{\phi}})\hat{\underline{\phi}}$ is substituted by the perturbed electric far field pattern:

$$\widetilde{\underline{E}}_0 = (1 + \epsilon\zeta_\theta)(\underline{E}_0, \hat{\underline{\theta}})\hat{\underline{\theta}} + (1 + \epsilon\zeta_\phi)(\underline{E}_0, \hat{\underline{\phi}})\hat{\underline{\phi}} \tag{4.16}$$

where $\epsilon > 0$, ζ_θ and ζ_ϕ are independent random numbers uniformly distributed in $[-1, 1]$ and $\hat{\underline{\rho}}$, $\hat{\underline{\theta}}$, $\hat{\underline{\phi}}$ are the unit vectors cartesian coordinate system to the spherical coordinate (ρ, θ, ϕ).

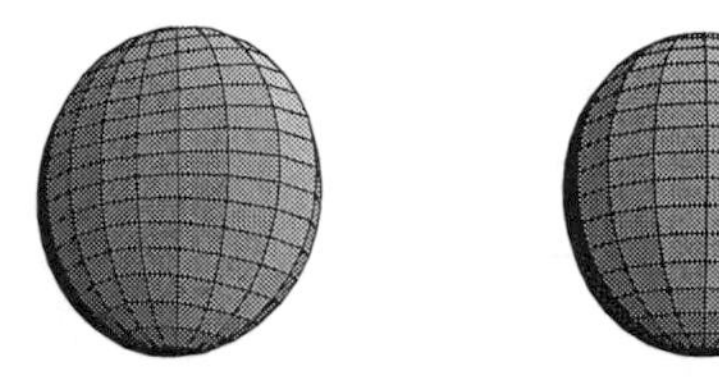

FIG. 3. *Original* *Reconstructed*
$\epsilon = 0.05, \;\; E^f_{L^2} \approx 0.033$

The set Ξ is taken as a uniform mesh on ∂B in the variables θ, ϕ, $\Omega_1 = \{3\}$ and finally Ω_2 is chosen such that the directions of propagation of the incoming plane wave are uniformly distributed on ∂B and for every direction of propagation considered two incoming waves with orthogonal polarization vectors are chosen.

The reconstruction is performed, as in the acoustic case, reformulating the inverse problem as an unconstrained optimization problem with 25 independent variables, i.e. the coefficients $\underline{c}$ of the obstacle shape when $L_\rho = 4$, see (4.1).

No a-priori information is used and the optimization problem is solved with the routine DUNLSJ [24] and the unit sphere is used as initial guess of the optimization procedure. The results obtained are shown in Fig. 3 and are taken from [20].

Example 4.4 We consider Problem 1.6 for the prolate ellipsoid:

$$\partial D = \left\{ (\rho, \theta, \phi) \in \mathbf{R}^3 \,\middle|\, \rho = \left[\sqrt{\sin^2\theta + \left(\tfrac{2}{3}\cos\theta\right)^2}\right]^{-1}, \; 0 \le \theta \le \pi,\, 0 \le \phi < 2\pi \right\} \tag{4.17}$$

with a piecewise constant impedance:

$$(4.18) \qquad \psi(\theta) = \begin{cases} -1, & 0 \le \theta < \frac{\pi}{3} \\ i, & \frac{\pi}{3} \le \theta < \frac{2\pi}{3} \\ -1, & \frac{2\pi}{3} \le \theta \le \pi. \end{cases}$$

The sets Ξ is taken as uniform mesh in ∂B, $\Omega_1 = \{1, 3, 5\}$ and Ω_2 as in the previuos example.

In [22] two conditions are derived, involving the impedance ψ and the shape ∂D of the obstacle. These conditions are linear in ψ, so that they allow us to obtain a new condition for ∂D independent of the impedance ψ. This latter condition is used to formulate an unconstrained optimization problem with 5 independent variables, i.e. the coefficients $\underline{c}$ of the obstacle shape, see (4.1) when $L_\rho = 4$ and assuming that the obstacle is x_3-axial symmetric. Finally, using the knowledge of the shape of the obstacle, and the original conditions the impedance of the obstacle is reconstructed by solving several local quadratic least squares optimization problems. In particular the i-th problem has χ_i as unknown for $i = 0, \ldots, N$, where $\chi_i = \chi(\theta_i)$, $\theta_i = i\frac{\pi}{N}$ and we have chosen $N = 36$. All the optimization problems are solved with the routine DUNLSF based on the Levenberg-Marquardt algorithm [24]. The results obtained are shown in Fig. 4 and are taken from [22].

For the reconstruction presented in Figures 1, 2, 3, 4 we have reported the following performance index. Let $h(\theta, \phi)$, $\tilde{h}(\theta, \phi)$, $0 \le \theta \le \pi$, $0 \le \phi < 2\pi$ be respectively a generic function and a numerical approximation of the same function, we define:

$$(4.19) \quad E^h_{L^2} = \sqrt{\frac{\sum_{i=1}^{N-1}\sum_{j=0}^{N} |h(\theta_i,\phi_j) - \tilde{h}(\theta_i,\phi_j)|^2 + |h(0,0) - \tilde{h}(0,0)|^2 + |h(\pi,0) - \tilde{h}(\pi,0)|^2}{\sum_{i=1}^{N-1}\sum_{j=0}^{N} |h(\theta_i,\phi_j)|^2 + |h(0,0)|^2 + |h(\pi,0)|^2}}$$

where $\theta_i = i\frac{\pi}{N}$, $\phi_j = j\frac{2\pi}{N}$, $i, j = 0, \ldots, N$. This notation is used also for functions $h(\theta)$, $\tilde{h}(\theta)$, $0 \le \theta \le \pi$ of only one variable. Finally we have chosen $N = 36$.

5. Conclusions. In this paper two methods to solve inverse acoustic and electromagnetic obstacle problems are described. These methods are presented in the hypothesis that the obstacle is a bounded simply connected domain with smooth boundary contained in a homogeneous medium that fills $\mathbf{R}^3 \setminus D$. Since this is a very simple setting for the inverse obstacle problem, it will be interesting to consider more general situations. A first step

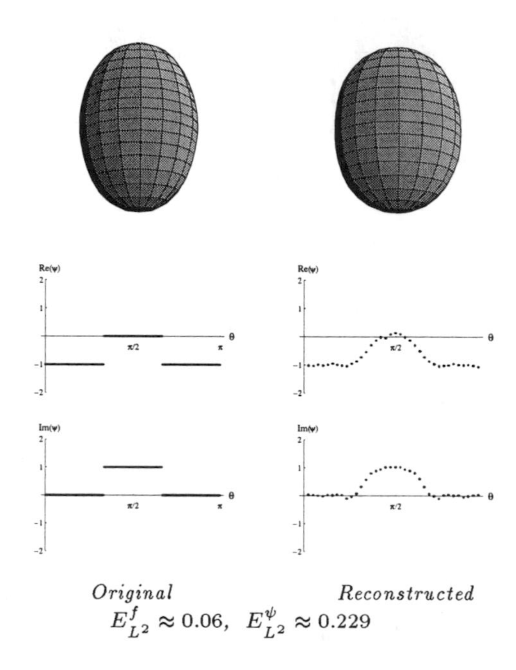

FIG. 4. *Original* *Reconstructed*

$E^f_{L^2} \approx 0.06, \quad E^{\psi}_{L^2} \approx 0.229$

should be to study algorithms for the reconstruction of obstacles D such that ∂D has multiscale corrugations, or obstacles with nonsmooth boundaries. Such reconstructions are possible when the characteristic lenght λ of the corrugations is at most of the same order of magnitude than the wavelenght of the incident radiation. A further step should be considering obstacle problems in layered media filling $\mathbf{R}^3 \setminus D$ or in layered media contained in regions different from $\mathbf{R}^3 \setminus D$ such as disturbed half-spaces. Finally another area of future research is given by the obstacle problem in randomly layered media. The increasing level of difficulty of these scattering problems suggests that starting from a given set of scattering data the reconstruction obtained with our methods degrades when the problem becomes more difficult, so that to keep the quality of the reconstructions presented here new procedures must be introduced. This effort must keep in mind that the efficiency of an algorithm depends on the architecture where the algorithm runs. For this reason recent developments of the hardware structures (parallel computers, multiprocessors) must be taken into account.

REFERENCES

[1] P. M. MORSE, K.V. INGARD, *Theoretical acoustics*, McGraw-Hill, Inc., New York, 1968.

[2] A. SOMMERFELD, *Partial differential equation in physics* , Academic Press, New York, 1964.

[3] D. COLTON, R. KRESS, *Integral equation methods in scattering theory*, J. Wiley & Sons Publ., New York, 1983.

[4] D. COLTON, R. KRESS, *Inverse acoustic and electromagnetic scattering theory*, Springer-Verlag, Berlin, 1993.

[5] J.A. STRATTON, *Electromagnetic theory*, McGraw-Hill Book Company, Inc., New York, 1941.

[6] J.B. KELLER, D. GIVOLI, *Exact non-reflecting boundary conditions*, J. of Comput. Phys., **82**, 1989, 172–192.

[7] P.C. WATERMAN, *New formulation of acoustic scattering*, J. Acoust. Soc. Amer., **45**, 1988, 1417–1429.

[8] G. KRISTENSSON, P. C. WATERMAN, *The T matrix for acoustic and electromagnetic scattering by circular disks*, J. Acoust. Soc. Amer., **72**, 1982, 1612–1625.

[9] B. PETERSON, S. STRÖM, *T matrix for electromagnetic scattering from an arbitrary number of scatterers and representations of E(3)*, Phys. Rev. D, **8**, 1973, 3661–3678.

[10] Z.L. WANG, L. HU, W. REN, *Multiple scattering of acoustic waves by a half-space of distributed discrete scatterers with modified T matrix approach*, Waves in Random Media, **4**, 1994, 369–375.

[11] P.M. MORSE, H. FESHBACH, *Methods of theoretical physics*, Part II, McGraw-Hill Book Company, Inc., New York, 1953.

[12] D.M. MILDER, *An improved formalism for wave scattering from rough surfaces*, J. Acoust. Soc. Amer., **89**, 1991, 529–541.

[13] L. MISICI, G. PACELLI, F. ZIRILLI, *A new formalism for wave scattering from a bounded obstacle*, to appear in Journal of the Acoustical Society of America.

[14] A.N. TIKHONOV, *On the solution of the incorrectly formulated problems and the regularization method*, Soviet. Math. Doklady, **4**, 1963, 1035–1038.

[15] D. COLTON, P. MONK, *A novel method for solving the inverse scattering problem for time-harmonic acoustic waves in the resonance region*, SIAM J. Appl. Math., **45**, 1985, 1039–1053.

[16] D. COLTON, P. MONK, *The numerical solution of the three dimensional inverse scattering problem for time harmonic acoustic waves*, SIAM, J. Sci. Stat. Comput., **8**, 1987, 278–291.

[17] F. ALUFFI-PENTINI, E. CAGLIOTI, L. MISICI, F. ZIRILLI, *A parallel algorithm for a three dimensional inverse acoustic scattering problem*, in "Parallel Computing: methods, algorithms and applications", D. J. Evans, C. Sutti Editors, IOP Publishing, Bristol, 1989, 193–200.

[18] L. MISICI, F. ZIRILLI, *Three dimensional inverse obstacle scattering for time harmonic acoustic waves: a numerical method*, SIAM, J. Sci. Stat. Comput., **15**, 1994, 1174–1189.

[19] P. MAPONI, L. MISICI, F. ZIRILLI, *A new method to reconstruct the boundary conditions of the Helmholtz equation*, in "Enviromental acoustics: International conference on theoretical and computational acoustics", vol. II, D. Lee, M. H. Schultz Editors, World Scientific Publisher Co., Singapore, 1994, 499–508.

[20] P. MAPONI, L. MISICI, F. ZIRILLI, *Three dimensional time harmonic inverse electromagnetic scattering*, in "Inverse problems in mathematical physics", L. Päivärinta, E. Somersalo Editors, Lecture Notes in Physics, Springer-Verlag, Berlin, **422**, 1993, 139–147.

[21] P. MAPONI, L. MISICI, F. ZIRILLI, *An inverse problem for the three dimensional vector Helmholtz equation for a perfectly conducting obstacle*, Computers Math. Applic., **22**, 1991, 137–146.

[22] P. Maponi, M.C. Recchioni, F. Zirilli, *Three dimensional time harmonic electromagnetic inverse scattering: the reconstrunction of the shape and the impedance of an obstacle*, Computers Math. Applic. **31**, 1996, 1–7.

[23] F. Aluffi-Pentini, V. Parisi, F. Zirilli, *Algorithm 667 SIGMA - A stochastic integration global minimization algorithm*, ACM Transactions on Mathematical Software, **14**, 1988, 366–380.

[24] The IMSL Libraries, IMSL Inc., P.O.Box 4605, Houston-Texas 77210-4605 (USA).

DESIGN OF 3D-REFLECTORS FOR NEAR FIELD AND FAR FIELD PROBLEMS

ANDREAS NEUBAUER*

Abstract. We report about a project concerning the computer-aided design of reflectors in $\mathbb{R}^3$ where the light coming from a point source is reflected in such a way that the illumination intensity distribution of the outgoing light on either a given plane or a given sphere can be prescribed. We derive a mathematical model and a numerical algorithm based on the iterative solution of a certain minimization problem. Finally, we present numerical results showing that the algorithm works in practice.

Key words. computer-aided design, near field problem, far field problem.

AMS(MOS) subject classifications. 65D17

1. Introduction. The problems we are dealing with are stated as follows: a point light source (lamp) with a known luminous intensity distribution is given. A reflector should be constructed in such a way that a prescribed illumination intensity is obtained

- either on a certain plane, called *near field problem*,
- or on a sphere centered at the lamp, where all reflected outgoing light rays are thought to emanate from the lamp, called *far field problem*. This problem can be interpreted as the limiting case of the near field problem when the plane to be illuminated moves out to infinity.

In both cases, the part of light going from the lamp to the illuminated plane or sphere directly (without being reflected) is known and needs not to be taken into consideration when calculating the reflector. It is substracted from the desired illumination intensity distribution at the very beginning.

The reflector should be a C^2-surface. This guarantees that the change in light radiation is continuously differentiable when deforming the reflector and that the change in the illumination intensity distribution is continuous.

The literature usually deals with the far field problem. Two-dimensional cases, i.e., cases where the light distribution of the lamp, the desired illumination distribution, and the reflector have either rotational or translational symmetry, were treated in [13,7]. Models for the three-dimensional case based on the solution of a nonlinear partial differential equation of Monge-Ampére type have been derived in [1,2,3,10,11]. Uniqueness results have been derived in [8,12].

All these models have in common that the relation *Incoming Ray - Outgoing Ray* is unique. This assumption is very often too restrictive in practice. A model and an algorithm based on the solution of a minimization problem, which does not need this assumption, have been derived in [9] for

* Institut für Mathematik, Universität Linz, A–4040 Linz, Austria.

the far field problem and in [6] for the near field problem .

The models derived in [9,6] are described in more detail in Sections 2 and 3. Numerical results are presented in Section 4 showing that the algorithms work in practice.

2. A model for the near field problem. In our model we assume that the light source is located at the origin of the coordinate system. The reflector is parametrized over a region G, using the independent variables (u, v). The incoming light ray $\boldsymbol{R_i}$ is normed in such a way that it is also the point on the reflector where this ray is reflected, i.e.,

$$\boldsymbol{R_i} = \begin{pmatrix} x(u,v) \\ y(u,v) \\ z(u,v) \end{pmatrix} .$$

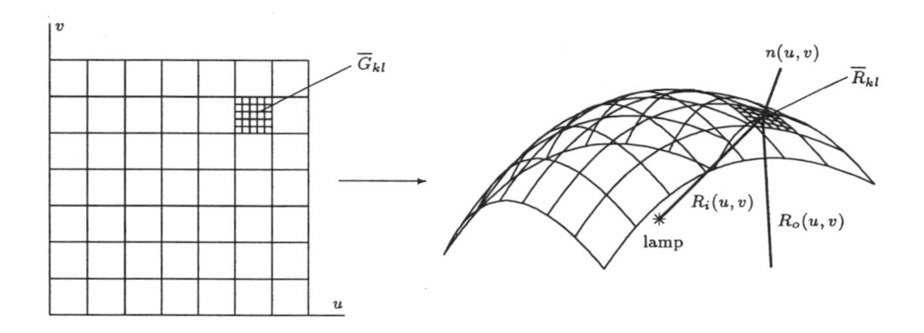

Figure 1: The parameter domain and the reflector.

For the reasons mentioned in the introduction, we assume that $\boldsymbol{R_i} \in C^2$. Moreover, we assume that we are dealing with a *geometric optics* problem, i.e. that the light rays are straight lines, and that the reflectors are ideal, so no scattering, diffraction or refraction of light will occur. Therefore, the reflected ray $\boldsymbol{R_o}$ obeys the law of reflection:

$$\boldsymbol{R_o} = \boldsymbol{R_i} - \frac{2\,\langle \boldsymbol{R_i}, \boldsymbol{n} \rangle}{||\boldsymbol{n}||^2}\,\boldsymbol{n} \, , \tag{2.1}$$

with

$$\boldsymbol{n} = \frac{\partial \boldsymbol{R_i}}{\partial u} \times \frac{\partial \boldsymbol{R_i}}{\partial v} = \begin{pmatrix} nx(u,v) \\ ny(u,v) \\ nz(u,v) \end{pmatrix} , \tag{2.2}$$

where $\langle \cdot, \cdot \rangle$ and $|| \cdot ||$ denote the Euclidean inner product and norm, respectively. See Figure 1 for an illustration.

In our model, no ray is allowed to be reflected two or even more times on the reflector. The illuminated plane is assumed to be parallel to the x-y-plane, i.e., $\overline{z} = -d$ $(d > 0)$. This special choice is not restrictive, since the reflector is represented as a function of (u, v). The coordinates $(\overline{x}, \overline{y})$ of the point, where a reflected ray hits the illuminated plane, are now calculated via the following expressions:

$$\overline{x} = x + \lambda \left(x - \frac{2 \langle \boldsymbol{R_i}, \boldsymbol{n} \rangle}{\|\boldsymbol{n}\|^2} nx\right),$$

$$\overline{y} = y + \lambda \left(y - \frac{2 \langle \boldsymbol{R_i}, \boldsymbol{n} \rangle}{\|\boldsymbol{n}\|^2} ny\right),$$

where

$$\lambda = - \frac{(z + d) \|\boldsymbol{n}\|^2}{z \|\boldsymbol{n}\|^2 - 2 \langle \boldsymbol{R_i}, \boldsymbol{n} \rangle nz}.$$

These expressions only make sense, if the following conditions hold:

- $z > -d$:
  The z-coordinates of all reflector points have to be larger than the z-coordinate of the illuminated plane. If this assumption is violated for one or even more reflector points, penetrations of the reflector and the illuminated plane will occur.
- $z - \dfrac{2 \langle \boldsymbol{R_i}, \boldsymbol{n} \rangle}{\|\boldsymbol{n}\|^2} nz \neq 0$: The z-coordinates of the reflected ray must not vanish. A violation of this condition means that the regarded ray is reflected parallel to the illuminated plane, and will never hit it.
- $\lambda > 0$: Because of the direction of the reflected ray $\boldsymbol{R_o}$, this parameter has to be positive. Otherwise, the reflected ray will be reflected away from the illuminated plane.

The basic requirement for the derivation of a model for the near field problem is the so called *light balance condition*. It says that the amount of light leaving the lamp in any solid angle increment $d\Omega$ should equal the amount of light reflected into the corresponding area increment dA in the illuminated plane or more precise: let ΔA be a sufficiently regular subarea on the illuminated plane and let ΔR be the largest subregion on the reflector such that for all incoming rays in ΔR the reflected outgoing rays hit points on the plane in ΔA. If $\Delta\Omega$ denotes the solid angle region and ΔG the appropriate subregion of the parameter domain G corresponding to ΔR, and if I and E denote the known luminous intensity distribution of the lamp up to a constant C and the prescribed illumination distribution on the plane, respectively, then the following equality holds:

$$C \int_{\Delta\Omega} I \, d\Omega = \int_{\Delta A} E \, dA \, . \tag{2.3}$$

Via the definition of surface integrals, we obtain for the integral on the left hand side

$$\int_{\Delta\Omega} I \, d\Omega = \int_{\Delta G} I(\gamma(u,v), \theta(u,v)) \, ||\boldsymbol{N}(u,v)|| \, d(u,v) \; . \tag{2.4}$$

The luminous intensity distribution I depends on $(\gamma(u,v), \theta(u,v))$, which are the spherical coordinates of the incoming ray $\boldsymbol{R_i}$. The connection between cartesian and spherical coordinates is chosen as follows:

$$\begin{aligned}
x &= r \sin\gamma \; \cos\theta \; , \\
y &= r \sin\gamma \; \sin\theta \; , \\
z &= r \cos\gamma \; , \\
r &= \sqrt{x^2+y^2+z^2} \; , \\
\gamma &= \text{arc}\cos\Big(\frac{z}{\sqrt{x^2+y^2+z^2}}\Big) \; , \quad \text{with } (x,y,z) \neq (0,0,0) \; , \\
\theta &= \begin{cases} \text{arc}\cos\big(\frac{x}{\sqrt{x^2+y^2}}\big) \; , & \text{if } y \geq 0 \; , \\ 2\pi - \text{arc}\cos\big(\frac{x}{\sqrt{x^2+y^2}}\big) \; , & \text{if } y < 0 \; , \end{cases} \quad \text{with } (x,y) \neq (0,0) \; .
\end{aligned}$$

For $\boldsymbol{N}$ in (2.4) the following holds:

$$\begin{aligned}
\boldsymbol{N} &= \left(\frac{\boldsymbol{R_i}}{||\boldsymbol{R_i}||}\right)_u \times \left(\frac{\boldsymbol{R_i}}{||\boldsymbol{R_i}||}\right)_v \\
&= \frac{\boldsymbol{R_{i_u}} ||\boldsymbol{R_i}|| - \boldsymbol{R_i} ||\boldsymbol{R_i}||_u}{||\boldsymbol{R_i}||^2} \times \frac{\boldsymbol{R_{i_v}} ||\boldsymbol{R_i}|| - \boldsymbol{R_i} ||\boldsymbol{R_i}||_v}{||\boldsymbol{R_i}||^2} \\
&= \frac{1}{||\boldsymbol{R_i}||^3} (||\boldsymbol{R_i}|| \boldsymbol{R_{i_u}} \times \boldsymbol{R_{i_v}} - ||\boldsymbol{R_i}||_u \boldsymbol{R_i} \times \boldsymbol{R_{i_v}} - ||\boldsymbol{R_i}||_v \boldsymbol{R_{i_u}} \times \boldsymbol{R_i}) \; ,
\end{aligned}$$

where the subindices u and v denote differentiation with respect to u and v, respectively. Together with

$$||\boldsymbol{R_i}||_u = \frac{\langle \boldsymbol{R_i}, \boldsymbol{R_{i_u}} \rangle}{||\boldsymbol{R_i}||} , \qquad ||\boldsymbol{R_i}||_v = \frac{\langle \boldsymbol{R_i}, \boldsymbol{R_{i_v}} \rangle}{||\boldsymbol{R_i}||}$$

and the formulae

$$\boldsymbol{a} \times \boldsymbol{b} = -\boldsymbol{b} \times \boldsymbol{a} \; , \qquad \boldsymbol{a} \times (\boldsymbol{b} \times \boldsymbol{c}) = \langle \boldsymbol{a}, \boldsymbol{b} \rangle \, \boldsymbol{c} - \langle \boldsymbol{a}, \boldsymbol{c} \rangle \, \boldsymbol{b} \; ,$$

which can be easily checked, the following formula follows:

$$\begin{aligned}
\boldsymbol{N} &= \frac{1}{||\boldsymbol{R_i}||^4} (||\boldsymbol{R_i}||^2 \boldsymbol{R_{i_u}} \times \boldsymbol{R_{i_v}} - \\
&\qquad - \boldsymbol{R_i} \times (\langle \boldsymbol{R_i}, \boldsymbol{R_{i_u}} \rangle \, \boldsymbol{R_{i_v}} - \langle \boldsymbol{R_i}, \boldsymbol{R_{i_v}} \rangle \, \boldsymbol{R_{i_u}})) \\
&= \frac{1}{||\boldsymbol{R_i}||^4} (||\boldsymbol{R_i}||^2 \boldsymbol{R_{i_u}} \times \boldsymbol{R_{i_v}} - \boldsymbol{R_i} \times (\boldsymbol{R_i} \times (\boldsymbol{R_{i_u}} \times \boldsymbol{R_{i_v}}))) \\
&= \frac{1}{||\boldsymbol{R_i}||^4} \langle \boldsymbol{R_i}, \boldsymbol{R_{i_u}} \times \boldsymbol{R_{i_v}} \rangle \, \boldsymbol{R_i}
\end{aligned}$$

Due to (2.2) and (2.4), (2.3) is equivalent to

$$C \int_{\Delta G} I(\gamma(u,v), \theta(u,v)) \frac{|\langle \boldsymbol{R_i}(u,v), \boldsymbol{n}(u,v)\rangle|}{||\boldsymbol{R_i}(u,v)||^3} \, d(u,v) = \int_{\Delta A} E(\overline{x}, \overline{y}) \, d(\overline{x}, \overline{y}) \, . \tag{2.5}$$

Note that the expression $\langle \boldsymbol{R_i}(u,v), \boldsymbol{n}(u,v)\rangle$ is not allowed to change sign. This condition has the physical interpretation that there are no unattainable points on the reflector.

Let us assume that the object to be illuminated in the plane $\overline{z} = -d$ is a subregion of the rectangle $[xl, xr] \times [yl, yr]$. Then the constant C in (2.5) may be determined by setting $\Delta A = [xl, xr] \times [yl, yr]$ and $\Delta G = G$. Since the total amount of reflected light only depends on the boundary of the reflector, C can be determined without knowing the reflector, if the boundary of the reflector is fixed as a constraint which is the case in many practical problems.

For deriving a numerical algorithm, the reflector has to be described by a finite number of variables. We choose a *bicubic B-Spline representation*. This representation makes sense for two reasons: the reflector surface is then two times continuously differentiable and changes in the spline coefficients have only local influence to the reflector surface.

Let $G = [0, nu] \times [0, nv]$ be a parameter domain, where nu and nv denote the number of subintervalls on the u- and v-axis, respectively. Then the B-Spline representation is of the following form:

$$\boldsymbol{R_i}(u,v) = \sum_{k=0}^{nu+2} \sum_{l=0}^{nv+2} \boldsymbol{cr_{kl}} N_k(u) N_l(v) \tag{2.6}$$

The spline coefficients $\boldsymbol{cr_{kl}} \in I\!R^3$ are the so called *de Boor control points* and the functions $N_k(u)$ and $N_l(v)$ are the well known cubic basis functions vanishing in $[0, nu] \backslash [k-3, k+1]$ and $[0, nv] \backslash [l-3, l+1]$, respectively (cf. [4]).

The aim is now to determine the coefficients $\boldsymbol{cr_{kl}}$ in equation (2.6) such that (2.5) holds best possible in the *least squares sense*. Obviously, equation (2.5) cannot be checked for all possible subsets ΔA of the illuminated plane. Since we have assumed that the object to be illuminated is a subregion of the rectangle $[xl, xr] \times [yl, yr]$, we use a uniform partition of this rectangle into $(2nx+1)(2ny+1)$ subareas; the region outside the rectangle, the *exterior*, is regarded as *one* further large subarea (see Figure 2 for an illustration of such a partition).

Thus, for all inner subareas A_{ij}, $i \in \{-nx, \ldots, nx\}$, $j \in \{-ny, \ldots, ny\}$ the following representation holds:

$$A_{ij} = [\overline{x}_i, \overline{x}_{i+1}] \times [\overline{y}_j, \overline{y}_{j+1}] \, , \tag{2.7}$$

where for all $k \in \{-nx, \ldots, nx+1\}$, $l \in \{-ny, \ldots, ny+1\}$

$$(\overline{x}_k, \overline{y}_l) = (\frac{k - \frac{1}{2}}{dx} + tx, \frac{l - \frac{1}{2}}{dy} + ty) \ ,$$

$$\begin{aligned} dx &= \frac{2nx+1}{xr - xl} \ , & tx &= \frac{xl + xr}{2} \ , \\ dy &= \frac{2ny+1}{yr - yl} \ , & ty &= \frac{yl + yr}{2} \ . \end{aligned} \tag{2.8}$$

(xl, yr) (xr, yr)

$(-nx, ny)$				(nx, ny)
		$(0,0)$		
$(-nx, -ny)$				$(nx, -ny)$

(xl, yl) (xr, yl)

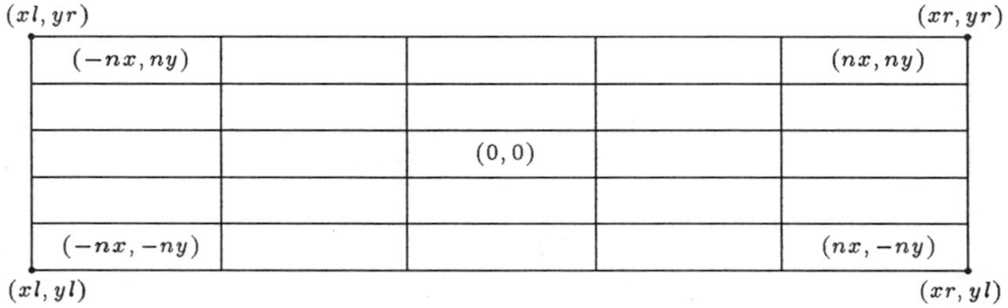

Figure 2: Partition for $nx = 2$ and $ny = 2$.

To solve (2.5) in the least squares sense means to determine the spline coefficients $\boldsymbol{cr_{kl}} \in I\!R^3$ of the reflector as the solution of the minimization problem

$$\sum_{i=-nx}^{nx} \sum_{j=-ny}^{ny} (lc_{ij}(\boldsymbol{cr_{kl}}) - ld_{ij})^2 + lco^2(\boldsymbol{cr_{kl}}) \rightarrow \text{Min!} \tag{2.9}$$

ld_{ij} denotes the luminous flux desired in the inner subarea A_{ij} and is given by

$$ld_{ij} = \int_{A_{ij}} E(\overline{x}, \overline{y}) \ d\overline{x} d\overline{y} \ .$$

$lc_{ij}(\boldsymbol{cr_{kl}})$ is the luminous flux calculated for the inner subarea A_{ij} and is given by

$$lc_{ij}(\boldsymbol{cr_{kl}}) = C \int_{G_{ij}} I(\gamma(u,v), \theta(u,v)) \frac{|\langle \boldsymbol{R_i}(u,v), \boldsymbol{n}(u,v) \rangle|}{||\boldsymbol{R_i}(u,v)||^3} \ d(u,v) \ ,$$

where G_{ij} is the largest region in G such that for all incoming rays from G_{ij} the reflected outgoing rays hit points in A_{ij}. $lco(\boldsymbol{cr_{kl}})$ denotes the sum of all light parts which belong to the exterior. The desired luminous flux for the exterior is always zero.

Since it is very complicated to calculate the regions G_{ij}, an *averaged version* $\overline{lc}_{ij}(\boldsymbol{cr_{kl}})$ is determined instead of $lc_{ij}(\boldsymbol{cr_{kl}})$ as follows:

We consider the subregions

$$\overline{G}_{kl} = [\frac{k-1}{m}, \frac{k}{m}] \times [\frac{l-1}{m}, \frac{l}{m}] , \quad k \in \{1, \ldots, nu \cdot m\}, \ l \in \{1, \ldots, nv \cdot m\} ,$$

with $m \in \mathbb{N}$. For each $\overline{G}_{kl}$ there exists a uniquely specified region $\overline{R}_{kl}$ on the reflector via (2.6). Via (2.5) the total light $\overline{lr}_{kl}$ being reflected from $\overline{R}_{kl}$ is given by

$$\overline{lr}_{kl} = C \int_{\overline{G}_{kl}} I(\gamma(u,v), \theta(u,v)) \frac{|\langle \boldsymbol{R_i}(u,v), \boldsymbol{n}(u,v) \rangle|}{||\boldsymbol{R_i}(u,v)||^3} \, dudv \, .$$

This integral is computed approximately using the Gaussian quadrature formula based on 4×4-nodes. For each of these 16 nodes the incoming and the outgoing rays are calculated via (2.1), (2.2) and the reflected rays are assigned the light value $\frac{\overline{lr}_{kl}}{16}$. For these 16 rays also the points where they hit the illuminated plane $\overline{z} = -d$ are calculated. Since we need in our algorithm that $\overline{lc}_{ij}(\boldsymbol{cr_{kl}})$ is continuously differentiable with respect to the spline coefficients $\boldsymbol{cr_{kl}}$, it is necessary to distribute the light values $\frac{\overline{lr}_{kl}}{16}$ onto the subregions A_{ij} defined by (2.7) in a differentiable way. This is achieved as follows (see Figure 3):

$$(2.10) \qquad \frac{\overline{lr}_{kl}}{16} \cdot \begin{cases} f(xx)f(yy) & \text{in } I := A_{i+1\,j+1} \\ f(xx)(1-f(yy)) & \text{in } II := A_{i\,j+1} \\ (1-f(xx))f(yy) & \text{in } III := A_{i+1\,j} \\ (1-f(xx))(1-f(yy)) & \text{in } IV := A_{ij} \end{cases}$$

where $f(s) = s^2(3-2s)$ for $s \in [0,1[$ and

$$\begin{aligned} i &= [dx(\overline{x} - tx)] \qquad & xx &= dx(\overline{x} - tx) - i \\ j &= [dy(\overline{y} - ty)] \qquad & yy &= dy(\overline{y} - ty) - j \end{aligned}$$

with dx, dy, tx, ty as in (2.8); the symbol [.] stands for the integer part of a real number.

The total amount of calculated luminous flux $\overline{lc}_{ij}(\boldsymbol{cr_{kl}})$ for an inner subarea A_{ij} is received by adding up all light value portions belonging to this subregion. If (i,j) in (2.10) does not belong to the set $\{-nx, \ldots, nx\} \times \{-ny, \ldots, ny\}$, then A_{ij} is part of the exterior; these light value portions are added up to get the luminous flux $\overline{lco}(\boldsymbol{cr_{kl}})$. The averaging effect of this calculation corresponds to the fact that in reality the lamp is not a point source, but extends over a small region, which has a smoothing effect on the light function anyway.

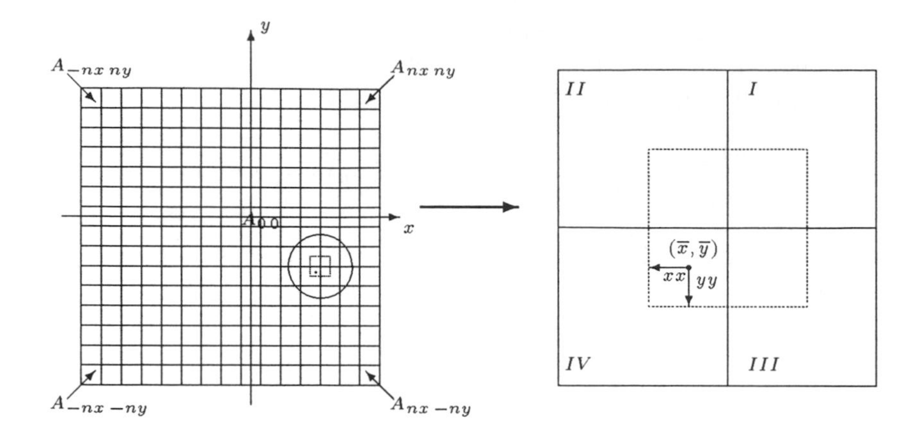

Figure 3: Distribution of the light value onto four subregions.

Instead of (2.9) we now solve the minimization problem

$$\sum_{i=-nx}^{nx} \sum_{j=-ny}^{ny} (\overline{lc}_{ij}(\boldsymbol{cr_{kl}}) - ld_{ij})w_{ij} + \overline{lco}^2(\boldsymbol{cr_{kl}})wo \to \text{Min!} \tag{2.11}$$

With the weights w_{ij} and wo the accuracy in the different subregions can be influenced. Together with (2.11) side conditions of the following type are used in practice:

- The boundary curve of the reflector can either be fixed or free.
- *Box contraints* are imposed on the B-spline coefficients to control the total size of the reflector.
- $\langle \boldsymbol{R_i}, \boldsymbol{n} \rangle$ is not allowed to change sign on the whole reflector surface.
- The conditions guaranteeing that all outgoing rays are reflected in the direction of the illuminated plane (see page 103) have to be satisfied.

Note that these conditions render problem (2.11) a constrained minimization problem; due to the last two conditions, the domain of admissible solution is usually not convex.

Since we are dealing with an inverse problem, stability questions arise. Note that we are not looking for some "ideal" reflector, but just for any reflector that minimizes (2.11). Therefore, we do not need stability for the coefficients $\boldsymbol{cr_{kl}}$, but only for the resulting light functions $\overline{lc}_{ij}(\boldsymbol{cr_{kl}})$ and $\overline{lco}(\boldsymbol{cr_{kl}})$. As mentioned above this stability is guaranteed in the C^1 sense. The stability question for the two dimensional far field problem was studied in [3].

3. A model for the far field problem. In this section we deal with the far field problem. It is derived in a completely similiar fashion as the model for the near field poblem. Therefore, we will only show the differences to the near field problem.

The interpretation of the *light balance condition* is now as follows: let ΔB be a sufficiently regular subregion on the unit sphere and let ΔS denote the corresponding subregion of $S := [0, \pi] \times [0, 2\pi]$ in spherical coordinates. Moreover, let ΔR be the largest subregion on the reflector such that for all incoming rays in ΔR the normed reflected outgoing rays are in ΔB. If the corresponding subregion in the parameter domain is denoted by ΔG and if I and $\tilde{I}$ denote the known luminous intensity distribution of the lamp up to a constant C and the prescribed illumination distribution on the unit sphere, respectively, then the following equality holds:

$$C \int_{\Delta R} I \, d\Omega_i = \int_{\Delta B} \tilde{I} \, d\Omega_o \, . \tag{3.1}$$

Via the definition of surface integrals we have already obtained an expression for the term on the left hand side (compare (2.5)). For the right hand side we obtain the expression:

$$\int_{\Delta \tilde{\Omega}} \tilde{I} \, d\tilde{\Omega} = \int_{\Delta S} \tilde{I}(\tilde{\gamma}, \tilde{\theta}) \|\Phi_{\tilde{\gamma}} \times \Phi_{\tilde{\theta}}\| \, d(\tilde{\gamma}, \tilde{\theta}) \, ,$$

where

$$\begin{array}{rcl} \Phi : S & \to & I\!R^3 \\ (\tilde{\gamma}, \tilde{\theta}) & \mapsto & \begin{pmatrix} \sin\tilde{\gamma}\cos\tilde{\theta} \\ \sin\tilde{\gamma}\sin\tilde{\theta} \\ \cos\tilde{\gamma} \end{pmatrix} . \end{array}$$

Together with (3.1),

$$\begin{aligned} \|\Phi_{\tilde{\gamma}} \times \Phi_{\tilde{\theta}}\| &= \left\| \begin{pmatrix} \cos\tilde{\gamma}\cos\tilde{\theta} \\ \cos\tilde{\gamma}\sin\tilde{\theta} \\ -\sin\tilde{\gamma} \end{pmatrix} \times \begin{pmatrix} -\sin\tilde{\gamma}\sin\tilde{\theta} \\ \sin\tilde{\gamma}\cos\tilde{\theta} \\ 0 \end{pmatrix} \right\| \\ &= \left\| \begin{pmatrix} \sin^2\tilde{\gamma}\cos\tilde{\theta} \\ \sin^2\tilde{\gamma}\sin\tilde{\theta} \\ \cos\tilde{\gamma}\sin\tilde{\gamma} \end{pmatrix} \right\| = |\sin\tilde{\gamma}| \end{aligned}$$

and due to the fact that $|\sin\tilde{\gamma}| = \sin\tilde{\gamma}$ for $\tilde{\gamma} \in [0, \pi]$, the light balance condition now reads as

$$\begin{aligned} C \int_{\Delta G} I(\gamma(u,v), \theta(u,v)) \frac{|\langle \boldsymbol{R_i}(u,v), \boldsymbol{n}(u,v)\rangle|}{\|\boldsymbol{R_i}(u,v)\|^3} \, d(u,v) = \\ \int_{\Delta S} \tilde{I}(\tilde{\gamma}, \tilde{\theta}) \sin\tilde{\gamma} \, d(\tilde{\gamma}, \tilde{\theta}) \, . \end{aligned} \tag{3.2}$$

Again the expression $\langle \boldsymbol{R_i}(u,v), \boldsymbol{n}(u,v)\rangle$ is not allowed to change sign. Choosing a *bicubic B-Spline representation* for the reflector as in Section 2, the aim is now again to determine the coefficients $\boldsymbol{cr_{kl}}$ in (2.6) so that

(3.2) holds best possible in the least squares sense. Obviously, one can not check (3.2) for all possible subsets ΔS of S. We have chosen a uniform partition of S into 3200 subsets:

$$(3.3)\quad S_{ij} := [\tilde{\gamma}_{i-1}, \tilde{\gamma}_i] \times [\tilde{\theta}_{j-1}, \tilde{\theta}_j]\,, \qquad i = 1, \ldots, 40\,,\ j = 1, \ldots, 80\,,$$

with

$$(3.4)\qquad (\tilde{\gamma}_i, \tilde{\theta}_j) := (i\frac{\pi}{40}, j\frac{\pi}{40}]\,, \qquad i = 0, \ldots, 40\,,\ j = 0, \ldots, 80\,.$$

As in Section 2, we will again calculate an averaged version for the left hand side in (3.2). To make it again differentiable, we now distribute the light values onto four neighbouring subregions in the following way (see Figure 4): let $(\tilde{\gamma}, \tilde{\theta})$ be the spherical coordinates of a reflected outgoing light ray and define

$$\begin{array}{lllllll} i & := & [\frac{1}{2} + \frac{\tilde{\gamma}}{h}] & , & g & := & \frac{\tilde{\gamma}}{h} - i \ \in [-\frac{1}{2}, \frac{1}{2})\,, \\ j & := & [\frac{1}{2} + \frac{\tilde{\theta}}{h}] & , & t & := & \frac{\tilde{\theta}}{h} - j \ \in [-\frac{1}{2}, \frac{1}{2})\,, \end{array}$$

where $h := \frac{\pi}{40}$ and $[\cdot]$ denotes the integral part of a real number. Then we distribute the appropriate light value onto the four subregions S_{ij}, $S_{i+1\,j}$, $S_{i\,j+1}$, $S_{i+1\,j+1}$ via

$$\frac{\overline{lr}_{kl}}{16} * \left\{ \begin{array}{llll} f(g)*f(t) & \text{in } I & := & S_{i+1\,j+1} \\ f(g)*(1-f(t)) & \text{in } II & := & S_{i+1\,j} \\ (1-f(g))*f(t) & \text{in } III & := & S_{i\,j+1} \\ (1-f(g))*(1-f(t)) & \text{in } IV & := & S_{ij} \end{array} \right.$$

where

$$f(s) := (1 + 3s - 4s^3)/2\,, \quad s \in [-\frac{1}{2}, \frac{1}{2})\,.$$

This definition makes sense, if we define $S_{0j} := S_{1j}$, $S_{41\,j} := S_{40\,j}$, $S_{i0} := S_{i\,80}$ and $S_{i\,81} := S_{i1}$ with S_{ij} as in (3.3), (3.4).

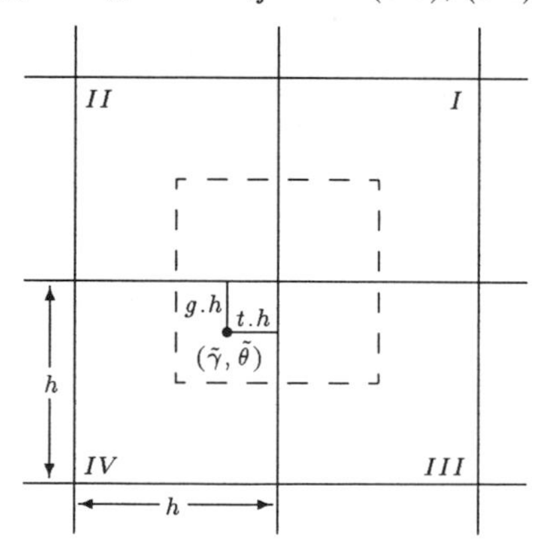

Figure 4: Distribution of the light value onto four subregions.

We now determine the spline coefficients $\boldsymbol{cr_{kl}} \in \mathbb{R}^3$ as the solution of the minimization problem

$$\sum_{i=1}^{40}\sum_{j=1}^{80}(\overline{lc}_{ij}(\boldsymbol{cr_{kl}}) - ld_{ij})^2\, w_{ij} \to \text{Min!}\ , \tag{3.5}$$

where $\overline{lc}_{ij}(\boldsymbol{cr_{kl}})$ is the averaged version obtained as described above (compare also Section 2) and

$$ld_{ij} = \int_{S_{ij}} \tilde{I}(\tilde{\gamma},\tilde{\theta}) \sin\tilde{\gamma}\; d(\tilde{\gamma},\tilde{\theta})\ .$$

Together with (3.5) side conditions of the following type are used in practice:

- The boundary curve of the reflector can either be fixed or free.
- *Box contraints* are imposed on the B-spline coefficients to control the total size of the reflector.
- $\langle \boldsymbol{R_i}, \boldsymbol{n}\rangle$ is not allowed to change sign on the whole reflector surface.

Note that, as in the section above, problem (3.5) together with these conditions is a constrained minimization problem.

4. Numerical aspects and results. The constrained minimization problems (2.11) and (3.5) are solved iteratively by a *projected conjugate gradient method*; by this we mean that the conjugate gradient direction is projected as in the steepest descent method, if a box constraint for the spline coefficients is attained. Since the changes in the coefficients are very small - these small changes have a large influence on the curvature of the reflector which is mainly influencing the light distribution on the illuminated plane or sphere, the projection of the conjugate gradient direction almost never occurs, and hence the algorithm is very effective in practice. The conjugate direction is calculated due to *Powell* and the line search is performed via *quadratic interpolation* (cf., e.g., [5]). The non-box constraints of the minimization problems (2.11) and (3.5), respectively, are controlled via the line search algorithm.

In each step of this iterative procedure the gradient of the functions in (2.11) and (3.5), respectively, has to be calculated. This is possible, if the known light distribution I of the lamp is continuously differentiable with respect to γ and θ. Note that then the light functions $\overline{lc}_{ij}(\boldsymbol{cr_{kl}})$ and $\overline{lco}(\boldsymbol{cr_{kl}})$ are continuously differentiable with respect to the spline coefficients $\boldsymbol{cr_{kl}}$. The exact calculation of this gradient is very complicated and messy, but the time spent for doing this calculation is worth it, since the CPU-time for one iteration can be reduced tremendously compared to the approach, where the gradient is approximated via a forward difference quotient. For instance, for the choice $nu = nv = 10$ or, equivalently, 507 variables the CPU-time was reduced by a factor of 75.

The convergence of our algorithm is tested for the near field problem with the following academic example: let $G = [0, nu] \times [0, nv]$ with $nu = nv = 10$; a plane test reflector is defined via

$$\begin{aligned} x(u,v) &= -10 + 2u \\ y(u,v) &= -10 + 2v \\ z(u,v) &= 1 \end{aligned}$$

The target area to be illuminated lies in the plane $\overline{z} = -30$ and has the extensions $[-320, 320] \times [-320, 320]$. If the luminous intensity distribution of the lamp is given by

$$I(\gamma, \theta) = \frac{1}{\cos^3 \gamma}$$

and if the desired illumination intensity distribution on the illuminated plane is given by

$$E = \frac{1}{1024} ,$$

then the academic reflector defined above satisfies the light balance condition (2.5) with $C = 1$.

Since the program calculates an averaged version of the light distribution, an error will occur even for the exact reflector. This error will tend to zero, if the number of subregions in the target area and the *hitting factor* tend to infinity. This factor denotes the average number of light rays per subregion in the illuminated plane and is calculated by

$$\frac{16 \; nu \; nv \; m^2}{(2nx + 1)(2ny + 1)} .$$

The larger the hitting factor, the better is the approximation to the exact illumination distribution. It turned out that this hitting factor should be at least ≥ 2.

To check the convergence of the algorithm, the initial reflector is chosen to be the following perturbation of the exact one:

$$\begin{aligned} x(u,v) &= -10 + 2u \\ y(u,v) &= -10 + 2v \\ z(u,v) &= 1 + 0.004 \; uv \; (10 - u)(10 - v) \end{aligned}$$

The "perturbed" reflector is shown in Figure 5.

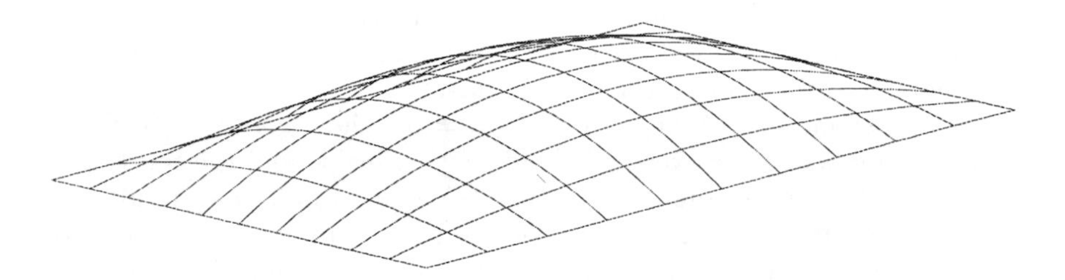

Figure 5: "Perturbation" of the plane reflector

For $m = 4$ and 1682 subregions on the target area to be illuminated this problem was solved on an Alpha-Station DEC 3000-600. The parameters C and the weights were put to 1. The CPU-time for 3000 iterations was about 9.2 hours. The final iterate was almost the exact plane reflector.

The l^2-errors for some iterations are given in the following table:

Number of Iterations	l^2 - Error
0	15229.3
1	11246.8
2	10858.0
5	8147.6
10	6008.8
100	2426.7
1000	194.5
2000	153.4
3000	122.9

Since there was no deviation from the desired illumination distribution in the exterior, in the next table only the percental average deviation inside is listed:

Number of Iterations	Deviation Inside
0	189.2 %
300	33.0 %
350	22.3 %
450	15.7 %
950	5.8 %
3000	3.8 %

We will now report about a practical example. The task was to calculate a so-called *wall-washer*, i.e., a reflector generating a uniform illumination intensity distribution on a plane. In the near field case, the area to be illuminated was assumed to have the extensions $[-80, 80] \times [-240, 240]$ in the plane $\overline{z} = -260$. The luminous intensity distribution of the lamp and the initial reflector were provided by an Austrian company that manufactures lighting fittings for industrial purposes.

When setting $m = 4$, $w_{ij} = 1$ and $wo = 0.1$ the algorithm calculates a reflector for which the average light deviation inside and the total light deviation outside were reduced by more than a factor of 3. This result was judged satisfactory by engineers of the company. Moreover, it took only 149 iterations, which corresponds to a CPU-time of five and a half hours on a VAX-Station 4000 VLC or, alternatively, 20 minutes on an Alpha-Station DEC 3000-600.

When choosing the weights in (2.11), special attention must be paid to an appropriate choice of wo, since the area covered by the reflected rays falling into the exterior is usually much larger than the inner subregions of the rectangle to be illuminated. If wo is chosen too large compared to the inner weights w_{ij}, all the light rays from the exterior are forced into the rectangle too fast with too less emphasis on the inner subregions. The algorithm might stop after a few iterations in a local minimum with a small deviation outside but a rather large deviation inside. Obviously, if wo is chosen too small, almost no light rays will be forced into the rectangle and hence there is not enough light in the interior to obtain a good approximation of the illumination intensity distribution. Since the size of the area covered by the rays in the exterior varies during the iteration process, it is too complicated to find a general rule for choosing wo.

How sensitively the algorithm responds to the choice of wo is illustrated in the following table by means of the wall-washer example:

wo	Number of Iterations	Deviation Inside	Deviation Outside
5.00	50	47.90 %	1.9 %
1.00	72	36.68 %	1.9 %
0.50	98	27.65 %	2.0 %
0.10	149	22.50 %	2.2 %
0.05	171	28.02 %	3.7 %
0.02	234	25.38 %	4.8 %
0.01	70	45.40 %	4.6 %

For the choice $wo = 0.1$ the deviation decreases uniformly inside and outside as can be seen in the next table:

Number of Iterations	Deviation Inside	Deviation Outside
0	71.3 %	7.3 %
50	33.2 %	3.5 %
100	25.7 %	3.1 %
149	22.5 %	2.2 %

The illuminated rectangle and the light distributions for the initial reflector and the final iterate are shown in Figures 6 and 7.

The shapes of the initial reflector and the final iterate are shown in Figures 8 and 9. Since the objective function is mostly influenced by the second derivative of the reflector, one has to look for changes in the curvature in comparing both figures.

The wall-washer example was also treated as a far field problem yielding similar satisfactory results than for the near field problem. To show that there is quite a difference in treating an example as a near field or as a far field problem we include the light distribution of the initial reflector in Figure 10 (compare Figure 6).

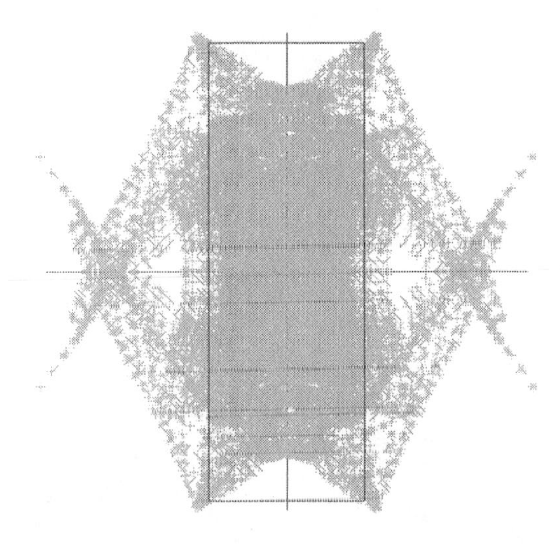

Figure 6: Light distribution for the initial reflector (near field)

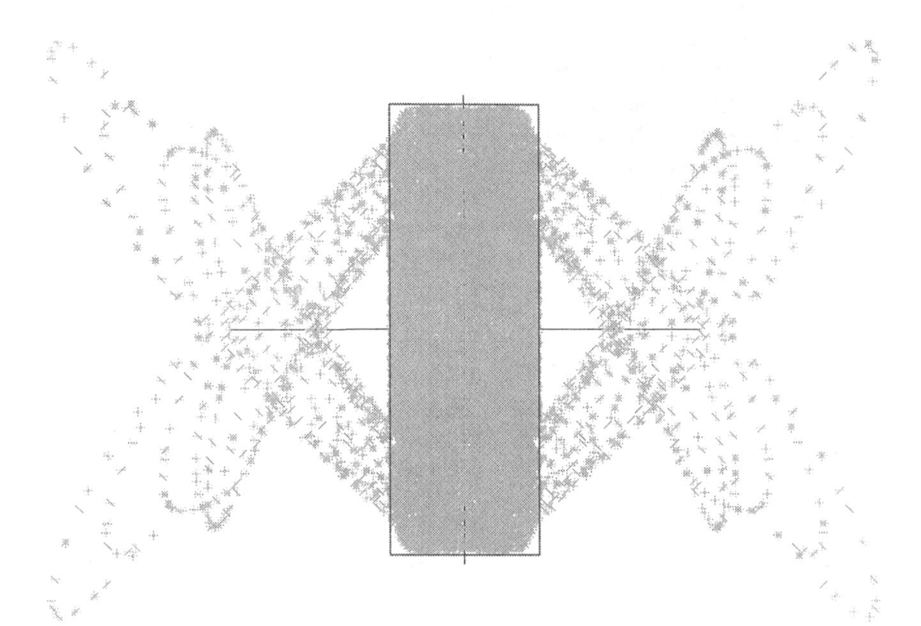

Figure 7: Light distribution for the final iterate (near field)

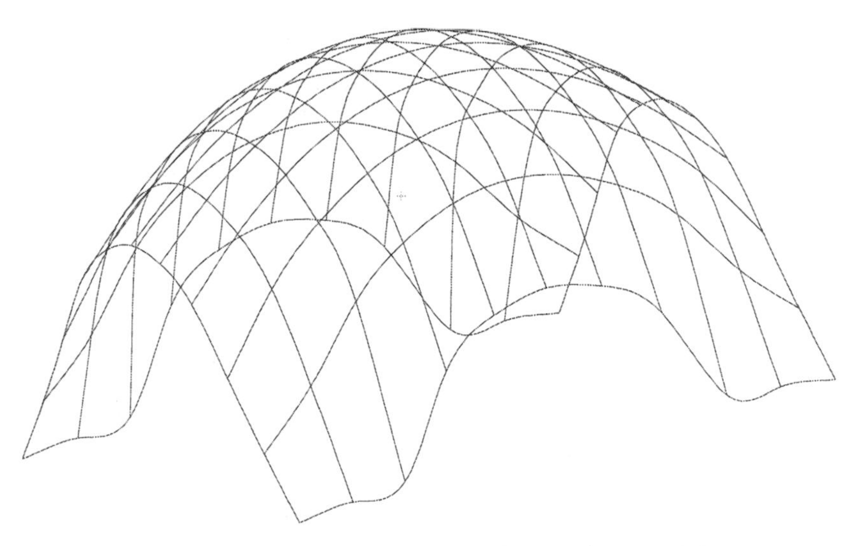

Figure 8: View of the initial reflector (near field)

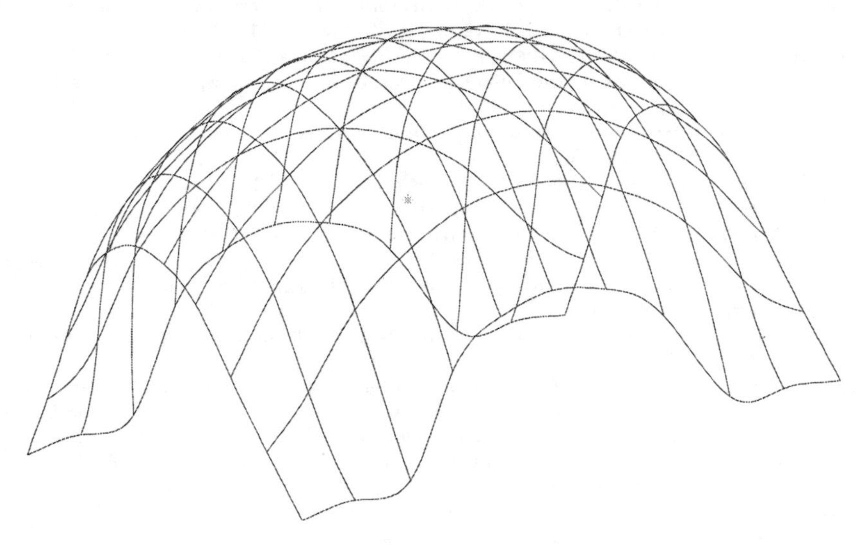

Figure 9: View of the final iterate (near field)

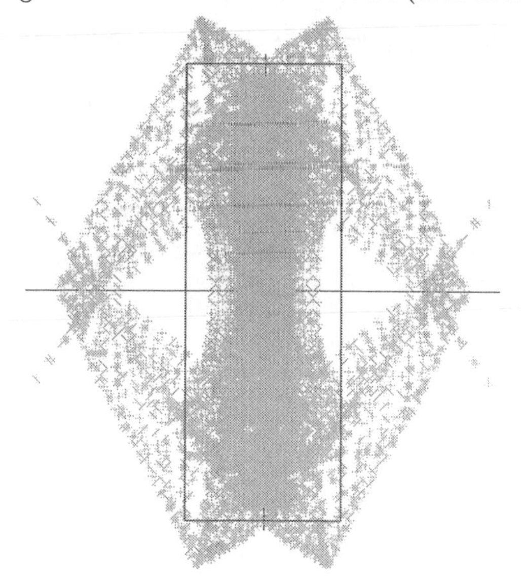

Figure 10: Light distribution for the initial reflector (far field)

REFERENCES

[1] F. Brickell, L. Marder and B.S. Westcott, *The geometrical optics design of reflectors using complex coordinates*, J. Phys. A: Math. Gen. 10 (1977), 246–260.

[2] F. Brickell and B.S. Westcott, *Reflector design for two-variable beam shaping in the hyperbolic case*, J. Phys. A: Math. Gen. 9, (1976), 113–128.

[3] H.W. Engl and A. Neubauer, *Reflector design as an inverse problem*, M. Heiliö (ed), Proceedings of the Fifth European Conference on Mathematics in Industry, Teubner, Stuttgart, 1991, 13–24.

[4] G. Farin, *Curves and Surfaces for Computer Aided Geometric Design*, Academic Press, Boston, 1990.

[5] R. Fletcher, *Practical Methods of Optimization*, Vol.1, Wiley, Chichester, 1980.

[6] C. König and A. Neubauer, *Design of a 3D-reflector for the near field problem: a nonlinear inverse problem*, Math. Engng. Ind. 5 (1996), 269–279.

[7] M. Maes, *Mathematical methods for 2D reflector design*, in: H.W. Engl, J. McLaughlin (eds.), Inverse Problems and Optimal Design in Industry, Teubner, Stuttgart, 1995, 123–146

[8] L. Marder, *Uniqueness in reflector mappings and the Monge-Ampere equation*, Proc. R. Soc. London, 1981.

[9] A. Neubauer, *The iterative solution of a nonlinear inverse problem from industry: design of reflectors*, in: P.J. Laurent, A.Le. Méhauté and L.L. Schumaker (eds), Curves and Surfaces in Geometric Design, AKPeters, Boston, 1994, 335–342.

[10] A.P. Norris and B.S. Westcott, *Computation of reflector surfaces for bivariate beamshaping in the elliptic case*, J. Phys. A: Math. Gen. 9 (1976), 2159–2169.

[11] B.S. Westcott, *Shaped Reflector Antenna Design*, Research Studies Press LTD, Letchworth, 1983.

[12] X.J. Wang, *On the design of a reflector antenna*, Inverse Problems 12 (1996), 351–375.

[13] W. Wolber, *Optimierung der Lichtlenkung von Leuchten*, Dissertation, Universität Karlsruhe, 1973.

OPTIMAL DIE SHAPE AND RAM VELOCITY DESIGN FOR METAL FORGING

LINDA D. SMITH*, JORDAN M. BERG†, AND JAMES C. MALAS III‡

Abstract. A primary objective in metal forming is designing the geometry of the workpiece and dies in order to achieve a part with a given shape and microstructure. This problem is usually handled by extensive trial and error–using simulations, or test forgings, or both–until an acceptable final result is obtained. The goal of this work is to apply optimization techniques to this problem. Expensive function evaluations make computational efficiency of prime importance. As a practical matter, off-the-shelf software is used for both process modeling and optimization. While this approach makes it possible to put together a working package relatively quickly, it brings several problems of its own. The algorithm is applied to a example of real current interest–the forging of an automobile engine valve from a high performance material–with some success.

1. Introduction. Forging is the the name given to a wide variety of metal forming processes. These have in common the following two characteristics: 1) The metal is shaped in its solid state, and 2) The shaping is through bulk deformation; that is, significant strains are induced throughout large portions of the workpiece. This is accomplished by squeezing the workpiece, or *billet*, between two or more shaped, hardened, *dies*. At least one of the dies is movable, driven by a *ram*. The initial billet shape is called the *preform*. Figure 1 is a schematic of the process.

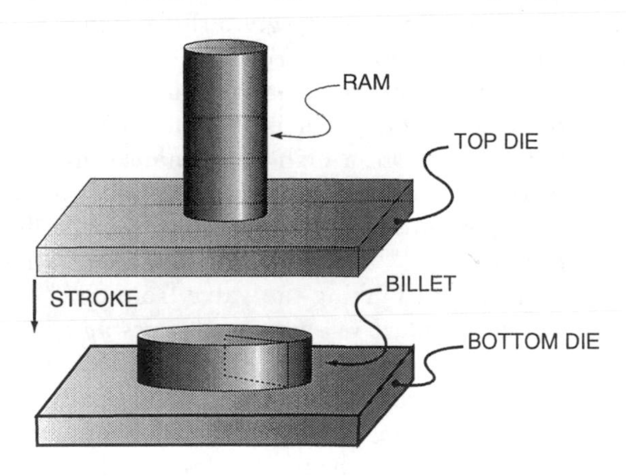

FIG. 1. *Forging Schematic.*

The large and distributed strains induced during forging are the key

* WL/FIGC-3 USAF Wright Laboratory Wright-Patterson AFB, OH 45433-7531.
† Department of Mechanical Engineering, Box 41021, Texas Tech University, Lubbock, TX 79409-1021.
‡ WL/MLIM USAF Wright Laboratory Wright-Patterson AFB, OH 45433.

to the value of the process, as compared to casting–where the resulting part is largely strain-free, or to machining–where the strain is extremely localized. The energy associated with the large plastic deformation can be used to transform the microstructure–either during the forging (*dynamic mechanisms*), or in post-forging heat treatment (*metadynamic* and *static* mechanisms). The most common use of a dynamic mechanism is certainly the control of yield strength via strain hardening. Static recrystallization is used to control grain size in rolled steel plates. Typically, higher strains correspond to more control over microstructure. However, this must be accomplished with some care, otherwise the material may fail. Thus, three considerations govern the design of a successful forging. The resulting shape should be as close to the final part as possible. Next, the deformation history must yield through dynamic, metadynamic, or static mechanisms an acceptable final microstructure. Finally, no cracks or voids are allowed in the finished part. The use of optimal design and control techniques in forging is relatively new. Little can be said about general approaches to the problem. Therefore this study concentrates on a specific problem. Forgings can be classified by the temperature of the billet during the forging. For *hot forging* the billet is heated to within sixty to seventy-five percent of its absolute melting point. This paper will consider only hot *isothermal forgings*, where the dies and the billet are enclosed by a furnace, and a desired forging temperature is maintained. The ram in such forgings will be driven by a hydraulic press. The combination of controlled material temperature and controlled ram velocity gives the designer maximum flexibility in choosing process parameters. For an account of design goals in other kinds of forging, and the use of optimization to achieve them, see [2].

The forging presented in this paper is that of a standard part–an automobile engine valve–made from a high-performance material. Figure 2 is a picture of the valve. The material, a Titanium Aluminide intermetallic, combines high temperature strength with low weight. The resulting valve would reduce valve "float" and improve engine performance at high rpm. The main problem with making the valve from TiAl is that, unlike steel, deformation must take place in a narrow *processing window* of strain, strain rate, and temperature. Furthermore, because this material retains its strength at high temperatures, the dies must be specially hardened, and care must be taken not to damage them during the forging. Lastly, economical manufacture requires that the production rate be maximized. This means that the total forging time must be as fast as possible. These three considerations have, to date, prevented a successful forging, and motivate this paper.

The valve is forged in two steps. Material stock is typically available from suppliers in simple forms, such as round or rectangular bars. The cheapest preform is a shape that can easily be cut from stock. For the valve, which has cylindrical symmetry, this means cutting sections from a round bar. Therefore, the first step of the forging is extruding the valve

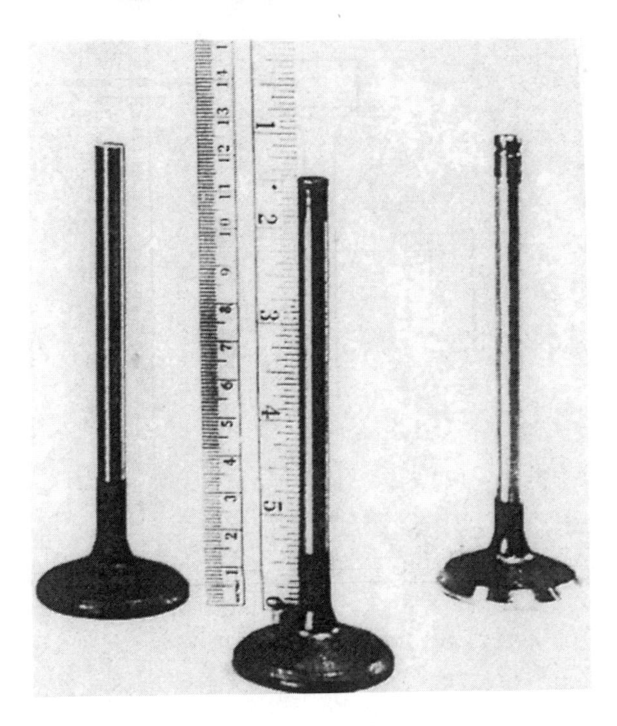

FIG. 2. *The Engine Valve.*

stem from the cylindrical preform. The second step is upsetting the valve head from the remaining material. Figure 3 shows the three stages. This paper considers only the extrusion step, from (a) to (b) in Fig. 3.

Section 2 describes the models that are available to characterize the forging. Section 3 develops the design variables, objectives, and constraints. Section 4 outlines the algorithm used to optimize the design. Section 5 presents the results.

2. The process model. A model of the forging process must consider two aspects. At the macroscopic level it will represent the continuum behavior of the billet. At the microscopic level it will describe the corresponding evolution of the material microstructure. These two length scales are coupled, with continuum states like strain, strain rate, and temperature driving microstructural transformations, and the microstructural states in turn determining the flow stress. Currently this coupling is all but ignored. The continuum thermo-mechanical fields are modeled by software packages using nonlinear finite- element methods (FEM). These simulations use flow stress data, tabulated as a function of strain, strain rate, and temperature, to account for microstructural evolution. When the appropriate material data is available, the nodal positions and velocities are accurate to within a percent. Models of microstructural transition are much less developed,

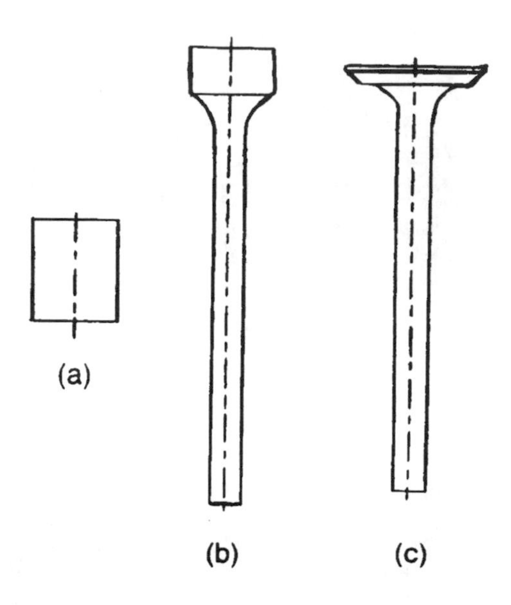

FIG. 3. *Two Step Valve Forging.*

especially for dynamic mechanisms. Transformations such as dynamic recrystallization remain poorly understood, even for common and industrially important materials such as steel.

This paper used the commercial FEM package *Antares* [9] as a flow model. *Antares* uses a rigid-viscoplastic approximation for the billet, and treats the dies as linear-elastic. The underlying PDEs, and general approach are described in [6]. The neglected elastic portion of the billet behavior is not significant for the high strains typical of forging. The savings in computation time are large, and that becomes far more important to an optimization than it is in solving the direct problem. The material itself enters *Antares* only as a database of flow stress information. *Antares* uses a Lagrangian formulation, with automatic mesh generation and rezoning. Output files can be generated that contain the nodal positions, velocities, and temperatures, as well as derived quantities, such as strain rate. It provides these capabilities for 2-D and 3-D part geometries.

The typical *Antares* user treats the simulation package as a cheaper, faster, alternative to a test forging. That is, the die and preform shapes are selected, a ram velocity profile, usually constant, or exponentially decaying, is specified, the forging temperature is fixed, and then the program is run, and the result evaluated. Simulation parameters such as time steps and mesh densities are tuned until the algorithm appears to be working reliably. If the result is unacceptable, the designer makes a change, and the process is repeated. Only the mesh rezoning is automated. Other simulation controls must be adjusted manually. This makes automation of the iterative design process–the goal of this work–very difficult. The simulation tends to be quite sensitive, and crashes easily. In practice this has meant constraining design variables to fairly narrow ranges, which are then relaxed or extended as appropriate. Numerical optimization will be faster and more reliable when more robust simulation methods are incorporated into the process model.

The TiAl intermetallic used in this study has been experimentally characterized in [3]. One means of presenting that information is a *processing map* [5]. Processing maps predict how a material will deform in terms of strain (how much the material deforms), strain rate (how fast the material deforms) and the temperature of the material. Figure 4 shows an example of a processing map for a TiAl alloy, at a fixed strain.

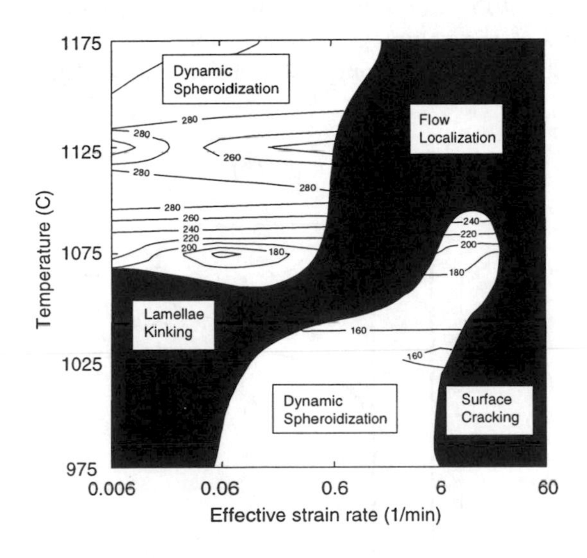

FIG. 4. *Processing Map for TiAl (following [3])*

The map shows "stable" regions, where the material may be deformed safely, and "unstable" regions where mechanical damage or undesirable microstructural mechanisms will render the material unfit for service. It is a hard constraint that the billet remains outside the unstable region during processing. Figure 4 shows the stable regions in white (with equivalent activation energy contours in the background), and unstable regions in black. Each region is labeled with its dominant microstructural mechanism.

3. Variables, objectives, and constraints.

3.1. The design variables. Recall that this paper considers only the first step of the valve forging, that is from (a) to (b) in Fig. 3. Figure 5 is a schematic of shape (b), with the design variables that define it shown. These variables have been chosen with the goal of achieving stable material flow throughout the forging. The most critical factor determining the maximum strain in the part is the extrusion ratio a, that is, the ratio of the billet area to the stem area. The extrusion ratio is therefore selected as a design variable. The most important factor in determining the peak strain rate (for a given ram velocity) is the shape of the transition region between the head and the stem. This region has been defined by its total length, l_{trans}, and the transition shape, which is given by a Bezier curve. The undetermined degrees of freedom in the Bezier curve are the coordinates of the second and third control points. These coordinates–in polar form $l_1, \theta_1, l_2, \theta_2$–define four design variables, and l_{trans} a fifth. There is no advantage to making the stem longer than its eventual length in shape (c), so the length of the stem in shape (b) is determined. Finally, the volume of material needed to form the head of the valve in shape (c) is fixed, and must correspond to the volume of material above the line $y = 0$ in Fig. 5. Therefore the above variables completely define shape (b).

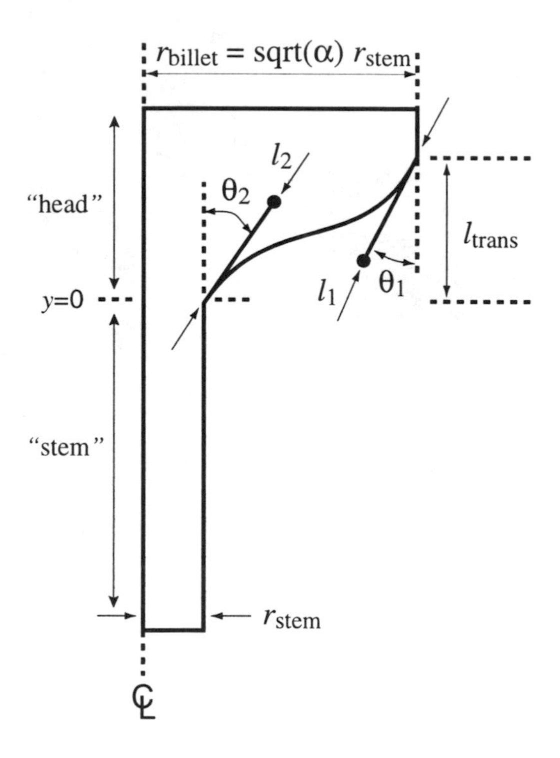

FIG. 5. *Valve Variable Definitions.*

Two issues remain. The first is that it may be desirable to allow the extrusion ratio to decrease by increasing the stem radius, rather than decreasing the billet radius. This would be the case if the billet was too slender for the subsequent upsetting. Therefore the stem radius is also allowed to vary. Of course, the extra material will have to be machined off, but if that extra step is costly it can be penalized in the cost function. The second issue is that the volume of revolution of the Bezier curve may exceed the head volume in shape (c). If the Bezier volume is less than the required volume there is no problem, the ram stroke is shortened, and the additional material necessary forms a cylinder over the transition region. There is no way, however, to drive the ram into the transition region. therefore the excess material must be trimed off, either before or after the final upsetting. This excess can also be penalized in the cost function, if desired. Note that the volume of revolution of the Bezier curve is a known function of the control points.

The variables just defined determine the billet radius, and the total volume, Thus they determine the preform shape. Clearly they also define the die shape. The only aspects of the forging remaining undefined is the ram velocity profile. The most natural way to specify this is in terms of ram velocity versus ram stroke. Stroke is the amount the ram has moved since first contacting the billet. This trajectory is parametrized by cubic splines. Two cubic splines, equally spaced in terms of stroke, were used. Five design variables are required to define this trajectory, namely the ram velocity at each control point, and the first and second derivatives at zero stroke. [8] describes a study using extremely simple geometries that suggested a two piece spline would suffice to describe the ram velocity. The results of the current paper suggest that for more complex shapes this is not adequate.

The design variables, and their relation to the other geometric parameters, are summarized below.

Design constants:

V_{head} : The volume needed to form the valve head in shape (c)
l_{stem} : The length of the valve stem in shape (c)

Design variables:

$x(0)$: Ram velocity at zero stroke
$x(1)$: Ram velocity at half stroke
$x(2)$: Ram velocity at full stroke
$x(3)$: First derivative of ram velocity at zero stroke
$x(4)$: Second derivative of ram velocity at zero stroke
$x(5)$: Extrusion Ratio (α)
$x(6)$: First control length of bezier (l_1)
$x(7)$: Second control length of bezier (l_2)
$x(8)$: First control angle of bezier (θ_1)
$x(9)$: Second control angle of bezier (θ_2)
$x(10)$: Stem radius (r_{stem})
$x(11)$: Transition length (l_{trans})

Derived quantities:

$$r_{\text{billet}} = r_{\text{stem}}\sqrt{\alpha}$$

$$V_{\text{billet}} = V_{\text{head}} + \pi r_{\text{stem}}^2 l_{\text{stem}} \quad \text{if } V_{\text{head}} > V_{\text{bezier}}$$

$$V_{\text{billet}} = V_{\text{bezier}} + \pi r_{\text{stem}}^2 l_{\text{stem}} \quad \text{if } V_{\text{head}} \leq V_{\text{bezier}}$$

$$h_{\text{billet}} = V_{\text{billet}}/(\pi r_{\text{stem}}^2 \alpha)$$

3.2. Cost function. There are many ways to express the goals of a successful forging in a cost function. For example, the boundaries of the stability regions, at the fixed forging temperature but varying with strain, may be used as constraints on the strain rate. Alternatively, a strain rate well inside a stable region, and corresponding to a desirable microstructural transformation mechanism, can be chosen. Then the distance from that strain rate, summed over all finite elements, and integrated over the entire forging, can be penalized. The second approach was used here, since it allows the designer to specify what makes a good forging, not just an acceptable one. The total forging time is also a natural quantity to minimize. The peak die stress, which must be kept below a maximum limit to avoid plastic deformation of the die, is more naturally framed as a constraint. This was not done for reasons of implementation, as described in the next section. That problem has been solved, and future efforts will treat the peak die stress as a constraint.

$$f(x) = \int_0^{t_{\text{final}}} \left(\frac{\sum_i (\dot{\epsilon}_{i_{\text{final}}} - \dot{\epsilon}_{\text{des}})^2 V_i}{\sum_i V_i} \right) + \sigma_{\max(\text{die})} + T_{\text{total}} \tag{3.1}$$

3.3. Constraints. The following constraints were applied to keep the ram velocity greater than a small positive value. This is necessary to prevent the constant time step from becoming too large for convergence. Each cubic spline takes its minimum either at the endpoint, or at an interior point satisfying a quadratic equation. The constraint is applied to the minimum point of each interval.

$$g_1(x) = 0.00001 - a_0 \leq 0 \tag{3.2a}$$

$$g_2(x) = 0.00001 - [a_1 + b_1(S_f - S) + c_1(S_f - S)^2 + d_0(S_f - S)^3] \leq 0 \tag{3.2b}$$

$$g_3(x) = 0.00001 - [a_0 + b_0(S_{\min}) + c_0(s_{\min})^2 + d_0(S_{\min})^3] \leq 0 \tag{3.2c}$$

$$g_4(x) = 0.00001 - [a_1 + b_1(S_{\min} - S) + c_1(S_{\min} - S)^2 + d_1(S_{\min} - S)^3] \leq 0 \tag{3.2d}$$

where

$$a_0 = x(0) \tag{3.2e}$$

$$b_0 = x(3) \tag{3.2f}$$

$$c_0 = x(4)/2.0 \tag{3.2g}$$

$$d_0 = [x(1) - x(0) - (x(3)S) - (x(4)/2.0)]/s^3 \tag{3.2h}$$

$$a_1 = x(1) \tag{3.2i}$$

$$b_1 = b_0 + 2c_0S + 3d_0S^2 \tag{3.2j}$$

$$c_1 = c_0 + 3d_0S \tag{3.2k}$$

$$d_1 = [x(2) - a_1 - b_1(S_f - S) = c_1(S_f - S)^2]/(S_f - S)^3 \tag{3.2l}$$

and S_f is full stroke, S is the location in stroke of the node of the cubic spline, and $S_{\min}$ is the stroke value at the minimum of that cubic polynomial.

4. Numerical optimization. The numerical approach to solving the die shape problem employs a gradient based nonlinear programming method to solve a constrained optimization problem. In particular, the Sequential Quadratic Programming (SQP) method is used to parametrically solve the die shape problem. SQP was chosen for its convergence properties. It also provides a numerically well conditioned approach to the optimization problem. SQP uses a technique where the search direction is found by solving a quadratic objective function and linear constraints. The search direction is found by creating a quadratic approximation to the objective function. The general problem set-up is as follows:

$$\text{Minimize: } Q(S) = f(x) + \nabla f(x)S + \frac{1}{2}S^T BS \tag{4.1a}$$

$$\text{Subject to: } \nabla g_j(x)S + \delta_j g_j(x) \leq 0;\ j = 1, m \tag{4.1b}$$

$$\nabla h_j(x)S + \overline{\delta}_j h_j(x) \leq 0;\ k = 1, l \tag{4.1c}$$

The design variable for this subproblem is S, the search direction. The matrix B is updated through iterations to approach the Hessian matrix. Once the search direction is found, a one dimensional search is performed with an exterior penalty function added.

$$\Phi = (x) + \sum_{j=1}^{m} u_j\{\max[0, g_j(x)]\} + \sum_{k=1}^{l} u_{m+k}|h_k(x)| \tag{4.1d}$$

where

$$x = x^{q-1} + \alpha S \tag{4.1e}$$

$$u_j = |\lambda_j|;\ j = 1, m+1 \text{ first iteration} \tag{4.1f}$$

$$u_j = \max[|\lambda_j|, \frac{1}{2}(u_j' + |\lambda_j|)] \text{ subsequent iterations} \tag{4.1g}$$

This algorithm includes several advantageous features. First, the Lagrangian is approximated and minimized using linear approximations to the constraints. The result, the search direction, is then used to minimize the augmented Lagrangian with no constraints. The update method for the B matrix maitains positive definiteness. All these combined provides an effective optimization tool. For further details see [7].

The SQP algorithm requires gradients with respect to each design variable for both the objective function and the constraints. Analytical computation of these gradients is advantageous since it requires a relatively low processing time. Unfortunately, analytical gradients were not available from *Antares*, and numerical approaches had to be used. The simplest numerical procedure is the forward finite difference analysis. The gradient is approximated by

$$f'(x) = \frac{f(x + \Delta x) - f(x)}{\Delta x} \tag{4.2}$$

where Δx defines the perturbation from the current problem solution. A one percent change in each design variable was used for the finite difference step.

Because the simulation controls of the *Antares* program have limited adaptive capability, the simulation is extremely sensitive to the ram velocity, and to the die geometry. Therefore, to prevent constant software crashes, each solution generated by the optimization routine must be feasible. To ensure this, this work used the optimization method FSQP [4]. It minimizes the maximum of a smooth objective function subject to smooth constraints. If the initial guess provided by the user is infeasible for some constraint, FSQP first works to generates a feasible point for these constraints; subsequently the successive iterates generated by FSQP all satisfy these constraints. The implementation of FSQP, CFSQP, had one serious drawback for the purposes of this study. It evaluates the cost function, and then evaluates the constraints one by one. Therefore, if the peak die stress is included as a constraint, CFSQP calls the FEM twice, once to evaluate the cost function, and once to evaluate the constraint. In fact, only one call is required. The use of numerical gradients exacerbated this problem. Because function evaluations are extremely expensive, this forced the peak die stress into the cost function. Since this study was carried out, the code has been modified, and now all functions generated by the FEM program are calculated simultaneously.

The finite element model was integrated into the optimization code as part of the objective function evaluation. It was also called when gradients of the objective function were required. The gradients were calculated for each design variable. A flow diagram of the algorithm is shown in Figure 6. The value num x in Figure 6 is the total number of design variables for the problem.

Although *Antares* is fairly speedy, as nonlinear FEM packages go, a single forward solution of a 2-D problem might take 30 minutes to an hour on a workstation. Since the ultimate goal of this research is to put optimal design tools into small forge shops, where the most powerful machine available is more likely to be a PC, these times will be significantly higher. Or, if 3-D geometries are considered, each forward solve could take days, even on a fast machine. Each function evaluation requires a complete Antares run. Each full gradient evaluation requires nine complete Antares runs! Therefore this application places a great premium on function evaluations. An optimization algorithm that kept function evaluations to an absolute minimum would probably be a better choice than SQP.

5. Results.

5.1. Design optimization. The differences between the forged part corresponding to the initial guess, and the forged part corresponding to the optimal solution are significant, as Fig. 7 clearly shows. The optimized shape has a much steeper slope for the Bezier curve and a significantly smaller head size. The nearly conical transition region evenly distributes both material strain rates and die stresses. The shape optimization aspects of this work are extremely encouraging.

Figure 8 compares the initial and final ram velocity profiles. The initial profile is a basically constant trajectory, at a relatively low ram velocity. The optimized ram velocity also starts slow, but climbs to nearly ten times its initial value before dropping again. Unfortunately this behavior is due more to the spline representation, and implementation, than to considerations of the forging. The ram velocity must be slow at the beginning, probably due to high stresses where the sharp corner of the billet contacts the die. As this corner wears down, the ideal ram velocity increases rapidly. Both remaining velocity control points are at their maximum value. But the maximum ram velocity itself, unlike the minimum ram velocity, is not constrained. Thus between the two control points the velocity increases as high as it can. In future work a consistent constraint will be applied to both the minimum and maximum velocities. Also an additional spline will be placed in the early portion of the forging. Alternatively, a pure ram velocity optimization, as described in [1] could be nested within the shape optimization loop.

Note that the solution has a shorter stroke than the initial guess. This is because the extrusion ratio and Bezier curve changed, and with them, the billet shape. In this case, the final head radius is smaller than the initial value, which tends to make the starting form longer, and increase the stroke, but this effect is more than offset by the slimmer stem radius and less volume in the transition region, of the final shape. Both these factors reduce the total volume of the preform, therefore tending to reduce its length, and so the total stroke.

The amount of time it takes to forge the entire valve was reduced from 118 minutes for the initial guess, to 11.5 minutes for the optimized solution. The decrease in time is directly attributable to the increase in ram velocity. The changes in die shape allowed the ram velocity to increase without overly increasing peak die stress or strain rate error.

The maximum die stress increased from the initial guess to the final optimized shape. The limit for the die stress is 35 ksi. The initial guess resulted in a maximum die stress of 11.17 ksi. The optimized result produced a maximum die stress of 24.71 ksi. This is largely due to the increase in ram velocity. The maximum die stress in the optimized forging is well within design limits. As mentioned above, future implementations will apply this term as a constraint.

The total integrated strain rate error was reduced from 40.92, for the initial guess, to 2.895 for the optimized result. As implemented, the reduction in total forging time plays a part in reducing this error term. This is not desirable, and in future work this term will be normalized by both the total forging time and the desired strain rate. Also, the square root will be taken, instead of using the square. Then the term has the meaning of normalized average strain rate error.

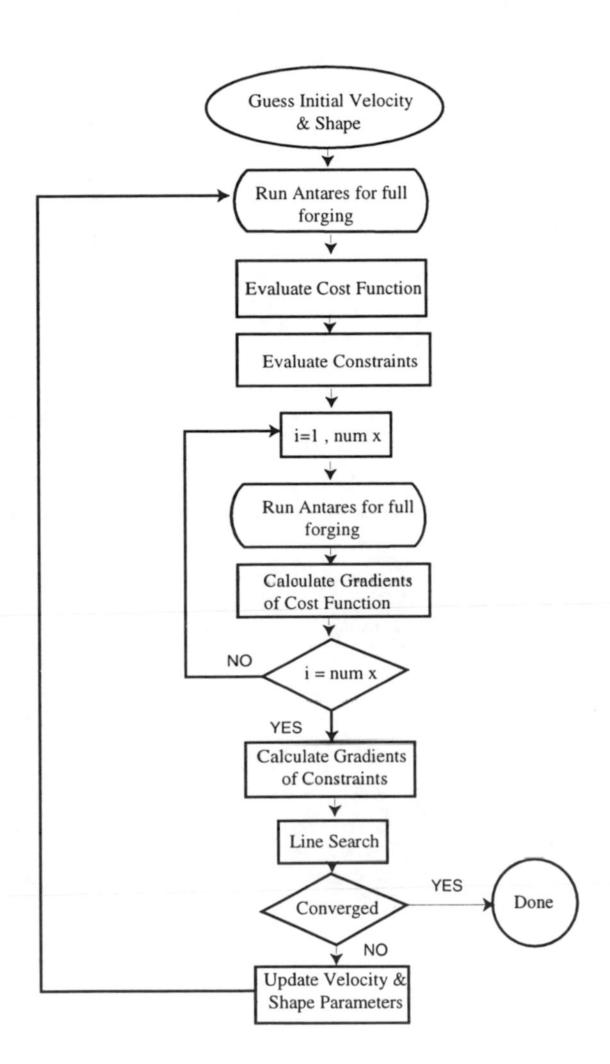

FIG. 6. *Optimization Flow Diagram.*

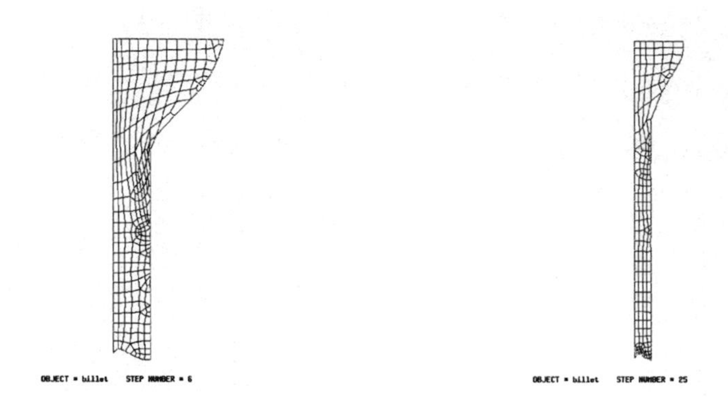

FIG. 7. *Forged shape corresponding to (a) initial guess (b) final solution.*

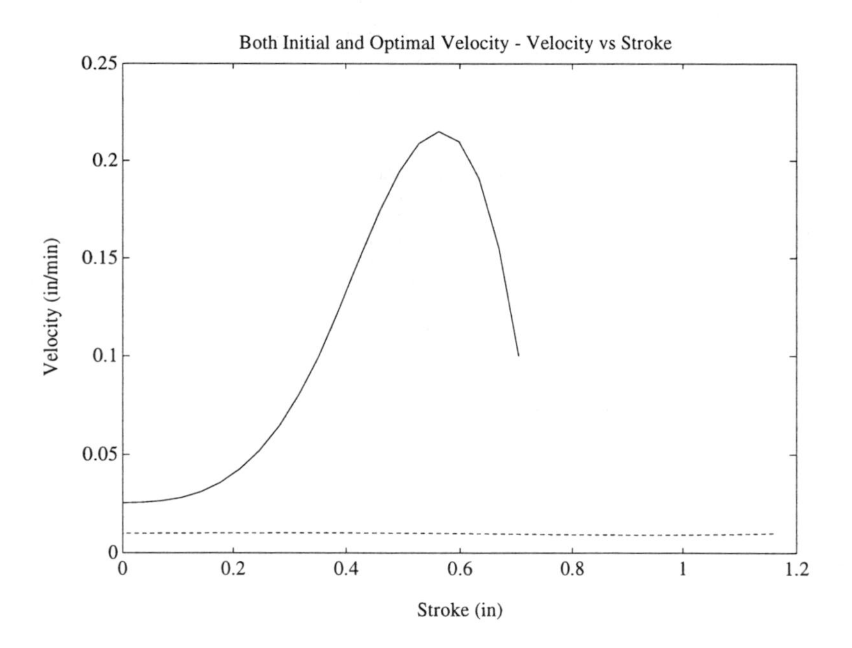

FIG. 8. *Initial (dashed) and Optimized (solid) Velocities*

5.2. Numerical performance. The convergence criteria used for this example was a change in objective function value of less than 0.1. After 46 function evaluations the run was terminated without convergence. The change in objective function was 0.3, which was judged to be sufficiently small. This decision was aided by the results of the design optimization, which appeared to be reasonable. This corresponded to three iterations through SQP. Three gradient calculations accounted for 36 of the function evaluations. The other 10 were done while in line search. The optimization took approximately 24 total hours to run on an SGI Iris Crimson workstation.

6. Conclusions. This paper considers a forging problem that is both representative of a wide class of interesting applications, and important in itself. The die and preform shapes, and the ram velocity profile, are parametrized, and then optimized. The cost function contains considerations of billet microstructure, die wear, and total forging time. This paper reports on a first attempt at formulating this design problem in an optimization context. It is clear after reviewing the results of this study that both the cost function and the ram velocity representation can be significantly improved. Despite this, the solution for die shape, in particular, appears excellent. It is anticipated that the next version of this work will improve the ram velocity calculations, while continuing to refine the shape optimization.

REFERENCES

[1] Berg, J., A. Chaudhary, and J. Malas, 1995. Open-loop control of a hot-forming process, in *Simulation of Materials Processing: Theory, Methods, and Applications: Proceedings of NUMIFORM '95*, S-F Shen and P.R. Dawson, editors, pp. 539–544, A.A. Balkema, Rotterdam.

[2] Boër, C.R., N. Rebelo, H. Rydstad, and G Schröder, 1986. *Process Modelling of Metal Forming and Thermomechanical Treatment*, Springer-Verlag, NY.

[3] Guillard, S. 1994. *High Temperature Micro-Morphological Stability of the* $(\alpha_2+\gamma)$ *Lamellar Structure in Titanium Aluminides*, Ph.D. Thesis, Materials Science and Engineering Dept., Clemson University.

[4] Lawrence, C., J. Zhou, and A. Tits, 1995. CFSQP *User's Manual Version 2.2*, University of Maryland, MD.

[5] Malas, J.C. and V. Seetharaman, 1992. "Using Material Behavior Models to Develop Process Control Strategies," *JOM*, Vol. 44, No. 6, pp. 8–13.

[6] Oh, S., 1984. "Finite element analysis of metal forming processes with arbitrarily shaped dies," *Int. J. Mech. Sci.* 24, 1982, 479–493.

[7] Schittkowski, K., 1985. NLPQP: A FORTRAN Subroutine Solving Nonlinear Programming Problems, Annals of Operation Research, 6 , pp.485-500.

[8] Smith, L., J. Berg, 1996. "Shape and Velocity Optimization for Hot Metal Forging," Proceedings of the 13th IFAC World Congress B, pp. 85–91.

[9] UES, Inc., 1993. *Antares User's Manual Version 4.0*, Dayton, OH.

EIGENVALUES IN OPTIMUM STRUCTURAL DESIGN

ULF TORBJÖRN RINGERTZ*

Abstract. Eigenvalues frequently appear in structural analysis. The most common cases are vibration frequencies and eigenvalues in the form of load magnitudes in structural stability analysis. In structural design optimization, the eigenvalues may appear either as objective function or as constraint functions. For example maximizing the eigenvalue representing the load magnitude subject to a constraint on structural weight.

Free vibration frequencies and load magnitudes in stability analysis are computed by solving large and sparse generalized symmetric eigenvalue problems. Eigenvalue constraints can therefore be represented using matrix inequalities as opposed to directly referring to the eigenvalues themselves. Since eigenvalues are nonsmooth functions of the design parameters, it is desirable to pose the constraints using matrix inequalities since this makes it possible to use a barrier transformation giving a smooth optimization formulation.

An overview of different structural design problems where eigenvalues appear as either constraints or objective function is given. In particular, it is described how barrier methods are useful for eigenvalue constraints. The more difficult case of unsymmetric matrices is also considered. An important application is structural optimization subject to aeroelasticity constraints which is briefly discussed.

Key words. Eigenvalue, structural optimization, matrix inequality, barrier method.

1. Introduction. Eigenvalue computation is an important part of structural analysis. Eigenvalues appear as vibration frequencies, buckling loads, and damping coefficients. The eigenvalue problem is usually posed in finite dimensional form where the matrices are obtained from a discretization, most frequently using a finite element formulation. The discretization must in most cases be quite detailed resulting in large and sparse eigenvalue problems where only a subset of the eigenvalues are of interest, for example those smallest in magnitude or having largest real part.

The optimal design of structures subject to eigenvalue constraints has a long history. The classical problem of finding the shape of a column that maximizes the buckling load for a given amount of material was studied by Lagrange and is still a current topic of research [1]. Many other structural optimization problems with eigenvalues have been studied, see [2,3,4] for extensive reviews.

The present paper gives an overview of different eigenvalue problems that may be part of structural design optimization. Then follows a description of some recent methods developed for the optimization of eigenvalues, both symmetric and unsymmetric matrices are considered. Finally, the important topic of imperfection sensitivity in design optimization of structures under nonconservative forces is discussed.

* Department of Aeronautics, Royal Institute of Technology, S-100 44 Stockholm, Sweden. This work was financially supported by the Swedish Research Council for Engineering Sciences (TFR).

2. Eigenvalues in structural analysis. Probably the most basic eigenvalue problem in structural analysis is the computation of free vibration frequencies obtained from solving

$$Kv - \omega^2 Mv = 0, \tag{2.1}$$

where K and M denotes the symmetric positive definite stiffness and mass matrices, v an eigenvector, and ω the vibration frequency. Since both matrices are symmetric positive definite, all eigenvalues ω^2 are real and positive such that all ω are real and well-defined.

Another important application is structural stability analysis. Consider the flat plate shown in Figure 2.1. The plate is subject to an in-plane

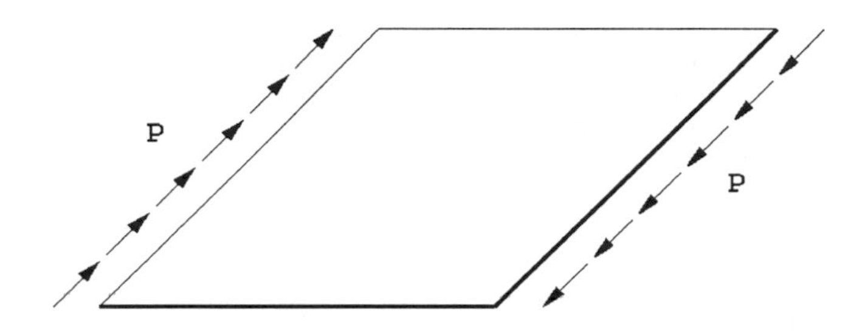

FIG. 2.1. *Flat plate with in-plane shear load*

shear load along two edges. The boundary conditions are such that the edges are free to rotate along the edges with applied load but cannot move in the directions perpendicular to the direction of the load. This is known as hinged boundary conditions. The other two edges are free.

For sufficiently small loads, the plate remains flat with only in-plane deformations. However, for a sufficiently large load, the potential energy ceases to have a minimum for the state with zero out-of-plane deformations and buckles as shown in Figure 2.2. The magnitude of the critical buckling load can be obtained as the smallest eigenvalue of

$$Kv - \lambda Fv = 0, \tag{2.2}$$

where F denotes the symmetric indefinite 'geometric stiffness' matrix and K the symmetric positive definite stiffness matrix. The eigenvalue λ is simply a scaling factor of the external load. The matrix sum $K - \lambda F$ represents an approximation to the Hessian of the potential energy. Since K is positive definite, the Hessian is positive definite for small λ representing a stable equilibrium state. However, for a sufficiently large load parameter λ, $K - \lambda F$ is indefinite making the equilibrium state with zero out-of-plane deformations unstable. The linear eigenvalue problem (2.2) only predicts

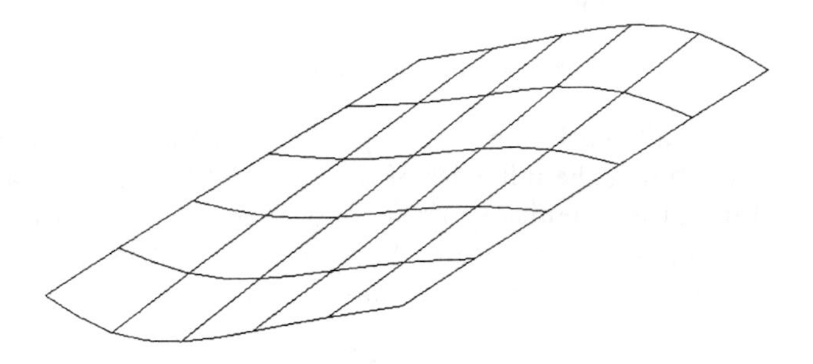

FIG. 2.2. *Postbuckled state of the plate*

the critical load, computing the actual deformations of the buckled plate, as shown in Figure 2.2, requires a geometrically nonlinear analysis [5].

2.1. Follower forces and unsymmetric matrices. The eigenvalue problems discussed in the previous section were both symmetric making computation of the smallest and most relevant eigenvalues straight-forward. However, in certain cases unsymmetric matrices appear in structural analysis.

Consider the two columns shown in Figure 2.3. The left column is

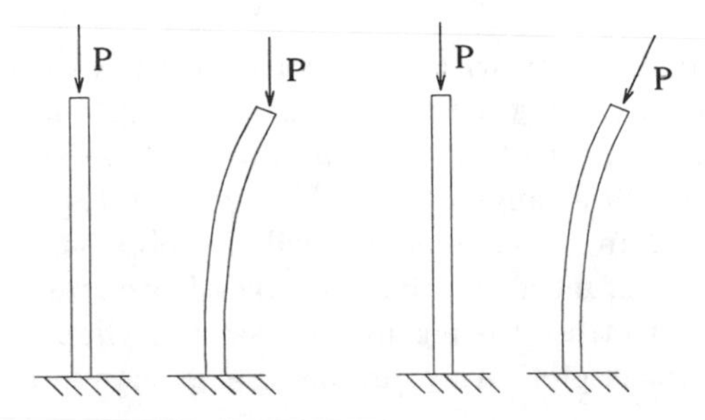

FIG. 2.3. *Two columns with different external loading*

subject to a conservative force which is such that it is independent of the deformation of the column. This makes it possible to compute the deformations by minimizing the potential energy of the column. The column is stable as long as the equilibrium state represents a minimum of the potential energy. The critical load is easily computed from the symmetric eigenvalue problem (2.2).

However, the right column is subject to a follower force meaning that the force remains tangential to the tip of the column. The column is often referred to as Beck's column [6]. There exists no potential function for this nonconservative mechanical system requiring the use of a different stability criterion. The stability of the column can in this case be investigated by considering the equations of motion

$$M\ddot{v} + D\dot{v} + Kv - pFv = 0, \tag{2.3}$$

where $\dot{v}$ denotes differentiation with respect to time, D the damping matrix, F the matrix defining the influence of the external force, and p a scalar parameter defining the magnitude of the load. The matrix F is unsymmetric but the damping matrix is in general symmetric positive definite.

Assuming that the deformations have the form

$$v = \tilde{v}\, e^{\lambda t}, \tag{2.4}$$

it is possible to derive the eigenvalue problem

$$[\lambda^2 M + \lambda D + K - pF]\, \tilde{v} = 0, \tag{2.5}$$

where λ is the eigenvalue and $\tilde{v} \in R^m$ the eigenvector. It is straight-forward to transform the quadratic eigenvalue problem (2.5) to a linear eigenvalue problem of the size $2m \times 2m$.

The column is considered stable if all eigenvalues λ have negative real part. The instability may be of two different forms; λ real and positive means strictly increasing deformation with time, and λ complex with positive real part means oscillations of increasing magnitude. These two instabilities are often referred to as *divergence* instability and *flutter* instability, respectively.

A significant difficulty with this nonconservative column is that the eigenvalue problem (2.5) is dependent on the magnitude p of the load as a parameter. This means that the column must be stable for all values of the parameter p in a range $(0, p_c)$ where p_c denotes the maximum load applied to the column. A series of eigenvalue problems, each problem with a different value of p, must be solved to investigate stability.

It is common to trace the eigenvalues as they change with the parameter p in a root-locus plot. An example plot is shown in Figure 2.4. The

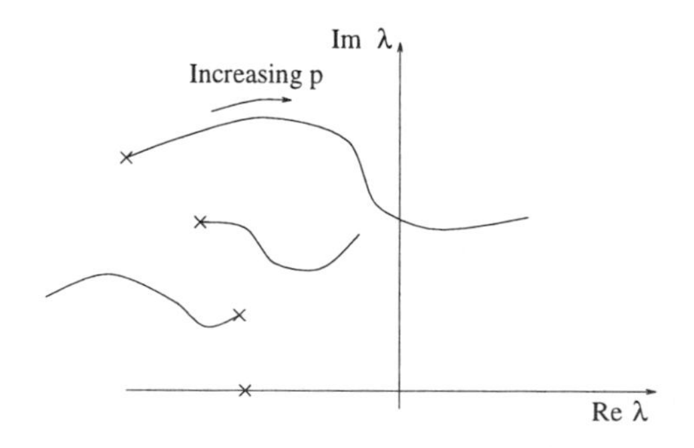

FIG. 2.4. *Root-locus plot*

eigenvalues all start out in the left halfplane for $p = 0$ but changes with increasing p. For a certain value of p, an eigenvalue crosses the imaginary axis and the column becomes unstable. This value of p defines the critical load of the structure.

2.2. Aeroelasticity. One of the most important applications of non-conservative forces is the aerodynamic forces on a wing structure. The interaction of the fluid flow and structural deformations can be analyzed using models of varying complexity. A reasonable approximate model is obtained by considering linear structural deformations and unsteady linear potential flow. The equations of motion are similar to those of the column (2.5) and are given by

$$\lambda^2 Mv + \lambda Dv + Kv - qF(\lambda)v = 0 \tag{2.6}$$

where the loading parameter q now denotes the dynamic pressure which is proportional to the square of the speed of flight. A significant difficulty is that the matrix of aerodynamic forces F depends nonlinearly on the eigenvalue λ. If compressible flow is assumed, the matrix F will also depend on the Mach number. This dependence on the eigenvalue causes the eigenvalue problem to be nonlinear which requires special solution techniques. However, it is in many cases possible to approximate $F(\lambda)$ with polynomial or rational functions making it possible to transform (2.6) to standard form.

The physical meaning of the two possible forms of instability is illustrated in Figure 2.5. The divergence instability appears when the wing

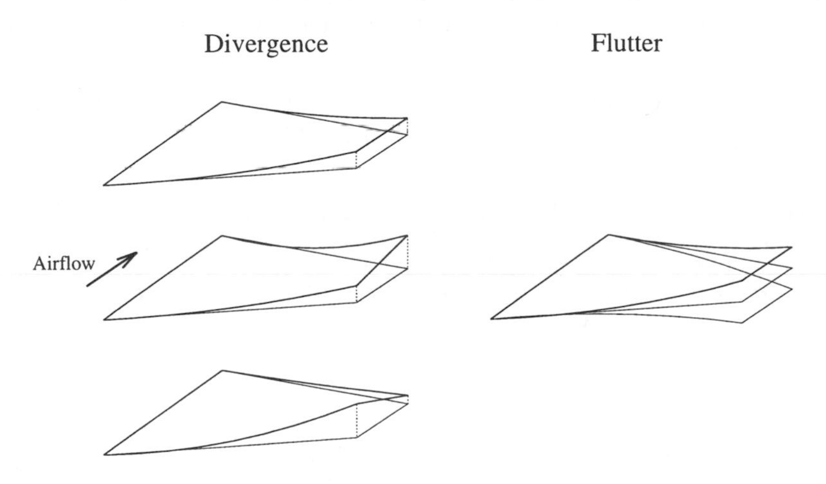

FIG. 2.5. *Aeroelastic instabilities*

bends such that the angle of attack increases, which increases the aerodynamic loads resulting in even larger deformations that ultimately lead to failure. This stability mode is especially important in the case of a forward-swept wing. The flutter instability is a vibration with deformations of increasing magnitude.

The design optimization of aircrafts wing subject to aeroelastic constraints is often referred to as *aeroelastic tailoring* [7,8].

3. Optimization of eigenvalues. Consider the simple model problem of maximizing the smallest eigenvalue of a symmetric matrix

$$\max_{x \in \Omega \subset R^n} \lambda_{min}(A(x)), \tag{3.1}$$

where A is a function of the variables x. It is well-known that this optimization problem is concave if $A(x)$ is a linear function of x. However, the problem can be quite difficult to solve since the smallest eigenvalue is a nonsmooth function of x. The nonsmoothness appears whenever there are coalescing eigenvalues.

It is sometimes assumed that a simple min-max transformation alleviates this difficulty. Introduce a slack variable z, giving

$$\max_{x,z} z \tag{3.2}$$

$$\lambda_i(x) \geq z \ , \quad i = 1...n, \tag{3.3}$$

the solution to (3.2)-(3.3) is obviously also a solution to (3.1). One may think that the nonsmoothness is now reduced to the ordering problem of keeping track of the eigenvalues and eigenvectors as a point where some eigenvalues coalesce is passed. Unfortunately, this is not true since the eigenvalues themselves are not smooth functions, as illustrated by the simple example given by Overton [9].

3.1. Barrier methods. If a logarithmic barrier transformation [10] is applied to the optimization problem (3.2)-(3.3) one obtains

$$\min_{x,z} -z - \mu \sum_{i=1}^{n} \log(\eta_i(x) - z). \tag{3.4}$$

This unconstrained function is then minimized for a decreasing sequence of barrier parameters $\{\mu^k\}$. It is straight-forward to show [11] that a solution to (3.4) converges to a solution of (3.2)-(3.3) under fairly mild conditions, when μ goes to zero.

The main advantage of minimizing the barrier function (3.4) is that it is a smooth function of the variables which is easily shown by using the following properties of the logarithm

$$\min_{x,z} -z - \mu \log \prod_{i=1}^{n} (\eta_i(x) - z) \;=\; \min_{x,z} -z - \mu \log \det(A(x) - zI). \tag{3.5}$$

The determinant is a smooth function of the matrix elements making the barrier function differentiable provided that the matrix elements themselves are smooth. This most useful property of the logarithmic barrier function has been used in several different applications, see for example [12,13,14,15].

The barrier function (3.4) can essentially be minimized using a standard method for unconstrained optimization provided that the linesearch is modified to ensure that the iterates stay strictly feasible. However, it has been found [15] that a modified Newton method using both directions of descent as well as directions of negative curvature is particularly useful.

In order to use a second-derivative method for the barrier function, it is necessary to compute both first and second derivatives of the function $\phi(x) = \log\det \hat{A}(x)$ which are given by the expressions

$$\frac{\partial \phi}{\partial x_j} = \operatorname{trace}(\hat{A}^{-1}\frac{\partial \hat{A}}{\partial x_j}) \tag{3.6}$$

$$\frac{\partial^2 \phi}{\partial x_j \partial x_k} = -\operatorname{trace}(\hat{A}^{-1}\frac{\partial \hat{A}}{\partial x_j}\hat{A}^{-1}\frac{\partial \hat{A}}{\partial x_k}) + \operatorname{trace}(\hat{A}^{-1}\frac{\partial^2 \hat{A}}{\partial x_j \partial x_k}) \tag{3.7}$$

The inverse is never formed explicitly, instead a Cholesky decomposition of $\hat{A}$ is used for computing the derivatives. The derivatives are somewhat difficult to compute since the matrix $\hat{A}(x)$ approaches a singular matrix as $\mu \to 0$. Some of the difficulties involved in computing the derivatives of the barrier function are discussed in [16].

3.2. Matrix inequalities. In using barrier for eigenvalue optimization, it is useful to formulate the eigenvalue constraints in terms of matrix inequalities. The notation $A \succeq B$ simply means that $A - B$ is positive semidefinite and the notation $A \succ B$ means that $A - B$ is positive definite.

Using this notation, the nonlinear constrained optimization problem

$$\min_{x \in \Omega} f(x) \tag{3.8}$$

$$A(x) \succeq 0 \tag{3.9}$$

simply means that the nonlinear function $f(x)$, representing for example structural weight, should be minimized subject to the constraint that the symmetric matrix $A(x)$ must be positive semidefinite.

The corresponding barrier minimization problem is

$$\min_{x \in \Omega} f(x) - \mu \log\det A(x). \tag{3.10}$$

Structural design constraints can easily be posed as matrix inequalities. The constraint that all vibration frequencies must be larger than a reference value ω_{ref} is posed as

$$K(x) - \omega_{\text{ref}}^2 M(x) \succeq 0 \tag{3.11}$$

where both the stiffness and mass matrices are assumed to depend on the design variable x.

The constraint that the buckling load parameter must be larger than a fixed value λ_c is posed as

$$K(x) - \lambda_c F(x) \succeq 0. \tag{3.12}$$

Several other formulations using matrix inequalities is possible, see for example Ringertz [15]. The special case of linear matrix inequalities is studied in detail in Boyd *et al.* [17].

4. Unsymmetric matrices. Optimization of eigenvalues of unsymmetric matrices is substantially more difficult than the symmetric case. Consider the simple matrix

$$A(x) = \begin{pmatrix} 0 & 1 \\ x & 0 \end{pmatrix}, \tag{4.1}$$

with the eigenvalues given by

$$\lambda = \pm\sqrt{x}. \tag{4.2}$$

The eigenvalues are clearly nonsmooth functions, actually not even Lipschitz continuous. This complicated behavior, with the additional difficulty that the eigenvalue problem often depends on a parameter, makes design optimization very difficult.

Consider the column problem governed by the equations of motion given by (2.3). The corresponding eigenvalue problem (2.5) can be put in standard form using the transformation $\tilde{u} = \lambda\tilde{v}$ which gives

$$\left[\begin{pmatrix} 0 & I \\ K - pF & D \end{pmatrix} - \lambda \begin{pmatrix} I & 0 \\ 0 & -M \end{pmatrix}\right] \begin{pmatrix} \tilde{u} \\ \tilde{v} \end{pmatrix} = 0. \tag{4.3}$$

If A is defined as

$$A = \begin{pmatrix} I & 0 \\ 0 & -M^{-1} \end{pmatrix} \begin{pmatrix} 0 & I \\ K - pF & D \end{pmatrix}, \tag{4.4}$$

the stability requirement becomes $\mathrm{Re}\{\lambda(A)\} \leq 0$. The matrix A depends on the designvariables x through the matrices M, K, and D. The force matrix F is in general independent of x.

4.1. Optimization. The optimization problem of finding the shape of the column that minimizes the structural weight $w(x)$ can be posed as

$$\min_{x \in \Omega} w(x) \tag{4.5}$$

$$\mathrm{Re}\ \lambda_i(A(p,x)) \leq 0\ , \quad i = 1, ..., m, \quad p \in (0, \hat{p}) \tag{4.6}$$

where x denotes the design variables describing the shape of the column and $\hat{p}$ the design load. This optimization problem is nonsmooth because of the eigenvalue constraints. The load parameter must also be discretized somehow since the stability constraint can not be enforced for all p in the range $(0, \hat{p})$.

4.2. Lyapunov's stability criterion. Because of the obvious difficulties involved with solving (4.5)-(4.6) it would be desirable to use a different optimization formulation. In particular, it would be useful if the stability constraint could be formulated using symmetric matrix inequalities making use of the barrier transformation possible.

One possibility is to use Lyapunov's stability criterion [17].

CRITERION 1 (LYAPUNOV). *There is a symmetric positive definite matrix $P \succ 0$ such that*

$$A^T P + PA \prec 0$$

if and only if all eigenvalues of A have Re $\lambda_i(A) < 0$.

This is precisely the condition that is needed for defining the stability constraint. There is also an alternative, possibly more useful statement of Lyapunov's criterion.

CRITERION 2 (LYAPUNOV). *The symmetric matrix P obtained by solving*

$$A^T P + PA + I = 0$$

is positive definite if and only if Re $\lambda_i(A) < 0$.

The optimization problem (4.5)-(4.6) may now be reformulated as

$$\min_{x \in \Omega, P_k} w(x) \tag{4.7}$$

$$- A^T(p_k, x)P_k - P_k A(p_k, x) \succ 0 \tag{4.8}$$

$$P_k \succ 0, \tag{4.9}$$

where the index k denotes the value of p for which stability is required. The stability constraint is enforced for a discrete index subset $\mathcal{K}$ of values p_k in the range $(0, \hat{p})$.

Alternatively using the second form of Lyapunov's criterion gives

$$\min_{x \in \Omega, P_k} w(x) \tag{4.10}$$

$$A^T(p_k, x)P_k + P_k A(p_k, x) + I = 0 \tag{4.11}$$

$$P_k \succ 0. \tag{4.12}$$

Note that both x and the elements of the symmetric matrix P_k are independent variables in the optimization problem. The second formulation (4.10)-(4.12) is somewhat more favorable since P_k is uniquely defined by (4.11) for given x whereas P_k may be multiplied by an arbitrary constant in (4.8)-(4.9).

4.3. Barrier function. The optimal design problem (4.10)-(4.12) is now posed in terms of symmetric matrix inequalities making it possible to use the logarithmic barrier transformation giving

$$\min_{x \in \Omega, P_k} w(x) + \mu \sum_{k \in \mathcal{K}} \log \det P_k \tag{4.13}$$

$$A^T(p_k, x)P_k + P_k A(p_k, x) + I = 0, \quad k \in \mathcal{K}. \tag{4.14}$$

Note that here the constraint on P being positive definite is applied to the inverse ($P^{-1} \succ 0$) since the Lyapunov equation forces the matrix P to become unbounded when the largest eigenvalue of A in real part approaches zero. To obtain a singularity of correct sign, the barrier transformation must be applied to the inverse. Using the properties of the logarithm gives

$$-\log \det P^{-1} = \log \det P. \tag{4.15}$$

The optimization problem (4.13)-(4.14) is large-scale even for moderate sized matrices A. The constraint Jacobian is sparse but the Hessian of the Lagrangian is dense because of the $\log \det P$ term. This problem could be solved with a sequential quadratic programming method where only the reduced Hessian is used and stored [15].

Another option is to use the equality constraints to eliminate P_k as an independent variable, giving

$$\min_{x \in \Omega} w(x) + \mu \sum_{k \in \mathcal{K}} \log \det P_k(x), \tag{4.16}$$

where $P_k(x)$ is obtained for given x by solving

$$A^T(p_k, x)P_k + P_k A(p_k, x) + I = 0. \tag{4.17}$$

This approach gives a problem with only x as independent variables. The elements of P_k can be considered as *state variables* which are obtained for given design x by solving the *state equations* (4.17).

The Lyapunov equation (4.17) represents a sparse linear system of equations in $(m(m+1))/2$ unknowns. Note that the system is sparse even if the $m \times m$ matrix A is dense. This system can be solved efficiently using a sparse LU factorization [18].

The barrier subproblem (4.16) is solved with a modified Newton algorithm [5] where both directions of descent and directions of negative curvature are used. The linesearch algorithm ensures that all the iterates stay feasible with respect to the stability constraints $P_k(x) \succ 0$.

5. Imperfection sensitivity. A significant difficulty in design optimization of structures subject to nonconservative forces is the sensitivity to small perturbations that sometimes is caused by optimization.

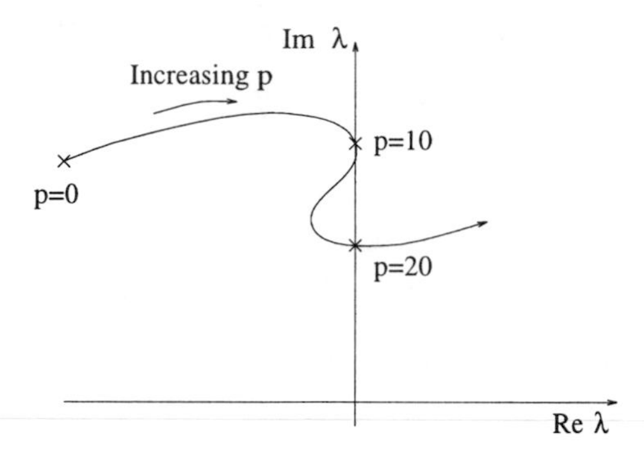

FIG. 5.1. *Root-locus plot of imperfection sensitive design*

Consider the root-locus plot shown in Figure 5.1 showing how one eigenvalue changes with the loading parameter p. The eigenvalue starts out well into the left half-plane and moves to the right for increasing p. For $p = 10$ the eigenvalue just touches the imaginary axis and then moves to the left again before finally crossing the imaginary axis for $p = 20$. This means that the stability constraint is active for $p = 10$ and $p = 20$. However, a small design change or perturbation of the data can move the eigenvalue into the unstable region for $p = 10$. This means that the critical load drops discontinuously from $p = 20$ to $p = 10$, clearly an undesirable situation.

To further illustrate this imperfection sensitivity, the column subject to a follower force in Figure 2.3 is optimized by finding the shape minimizing the weight subject to being stable for the loads $p = \{0.4, 0.8, 1.2, ..., 20.0\}$. The column is discretized by 10 finite elements with piecewise linear cross-sectional area along the length of the column. The damping model is so-called proportional damping with $D = \alpha K + \beta M$ where $\alpha = \beta = 0.1$, see [19] for further details.

The root-locus plot of the final optimal design is shown in Figure 5.2. Clearly, the stability constraint is active for two different loads, namely $p = 11.6$ and $p = 20$. For $p = 20$ two eigenvalues are critical (shown with a $\times$ symbol) and for $p = 11.6$ one eigenvalue is critical. A small perturbation may make the column unstable for $p \approx 11.6$.

It would be highly desirable to constrain the column to have a certain stability margin for the loads lower than the design load $p = 20$. One could, for example, require the eigenvalues to have real part less than some negative constant for the lower loads. The question is whether the magnitude of the eigenvalues is a good measure of nearness to instability.

An alternative measure of nearness to instability is to compute the smallest perturbation of a stable matrix making it unstable [20]. Consider the minimum norm optimization problem

$$\min_{E} ||E||_2 \tag{5.1}$$

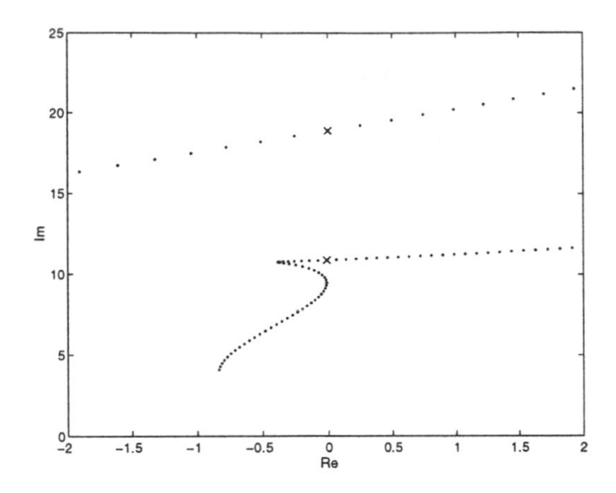

FIG. 5.2. *Root-locus plot for the optimized column*

$$A + E \text{ unstable} \tag{5.2}$$

where E is the perturbation matrix. The minimum E can, assuming it is complex, be computed using an algorithm suggested by Byers [21]. If A is real symmetric $||E||_2$ is simply the magnitude of the smallest eigenvalue, but if A is unsymmetric $||E||$ may be much smaller.

Table 5.1 shows the eigenvalues and minimum perturbations for the optimized column and some different values of the load p. Clearly, the

TABLE 5.1
Eigenvalues and perturbations for different p

p	$\max_i \mathrm{Re}\{\lambda_i\}$	$\|\|E^*\|\|$
4	-0.605	$0.261 \cdot 10^{-2}$
8	-0.199	$0.204 \cdot 10^{-3}$
11.6	-0.014	$0.523 \cdot 10^{-5}$
16	-0.370	$0.547 \cdot 10^{-4}$
20	-0.001	$0.157 \cdot 10^{-6}$

minimum perturbations are much smaller than the magnitude of the eigenvalues.

To further demonstrate the sometimes significant imperfections sensitivity, the eigenvalue with maximum real part and the minimum perturbation are shown as functions of p in Figures 5.3 and 5.4. Note the logarithmic scale in Figure 5.4. Another interesting observation is the the eigenvalue is not even a good *relative* measure of nearness to instability. The eigenvalue is smaller in magnitude for $p = 8$ than it is for $p = 16$, but in terms of minimum norm perturbation the column is closer to being unstable for $p = 16$ than it is for $p = 8$.

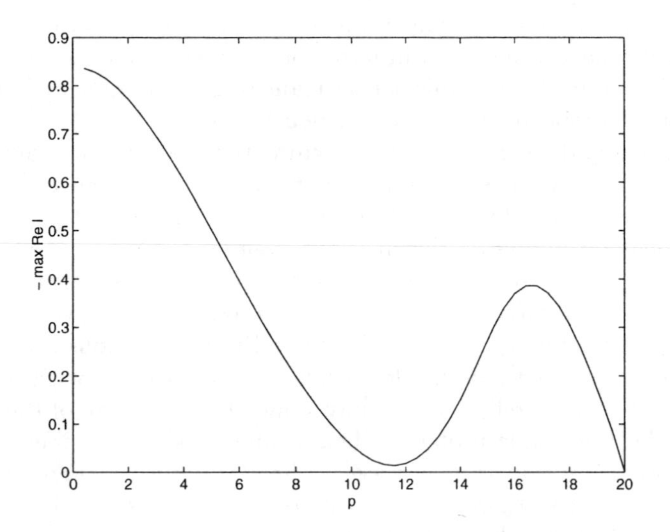

FIG. 5.3. *The eigenvalue with maximum real part versus the load p*

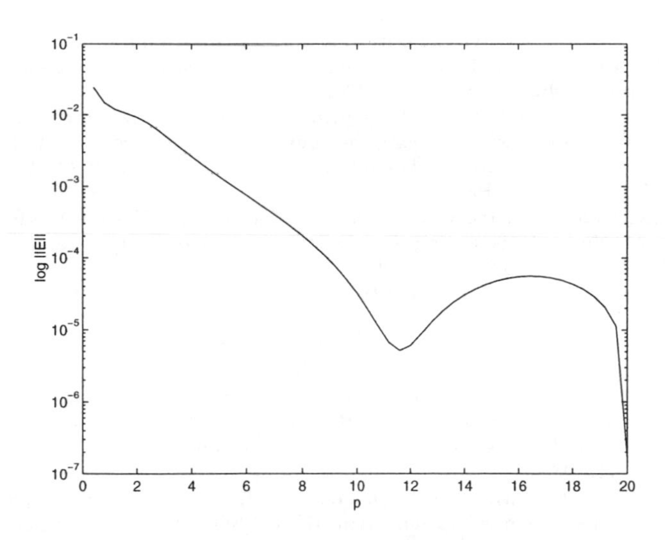

FIG. 5.4. *The minimum norm perturbation versus the load p*

6. Conclusions. Structural optimization subject to eigenvalue constraints may appear rather difficult because of the apparent nonsmoothness of the constraint functions if posed in terms of eigenvalues. However, if the constraint can be posed in terms of symmetric matrix inequalities it is possible to use a logarithmic barrier transformation resulting in a smooth optimization problem that can be efficiently solved.

Optimizing the eigenvalues of unsymmetric matrices may appear even more difficult since the eigenvalues are in this case not even Lipschitz continuous. Fortunately, Lyapunov's stability criterion makes it possible to use matrix inequalities and barrier methods even for the unsymmetric case.

Perhaps the most serious difficulty remaining is the sensitivity to imperfections. The magnitude of the eigenvalues is not a good measure of nearness to instability, as clearly illustrated in the previous section. Consequently, it would be desirable to pose the constraint with a stability margin that is more relevant, possibly some norm measure of the matrices forming the eigenvalue problem. This would make it possible to find an optimal structure that is guaranteed to be stable even if there are perturbations present less than a prescribed tolerance. However, it is not clear how these tolerances should be chosen.

REFERENCES

[1] S. J. Cox and M. L. Overton. The optimal design of columns against buckling. *SIAM J. on Mathematical Analysis*, 23:287–325, 1992.

[2] N. Olhoff and J. E. Taylor. On structural optimization. *J. Appl. Mech.*, 50:1139–1151, 1983.

[3] Z. Mroz. Design sensitivity of critical loads and vibration frequencies of nonlinear structures. In G. I. N. Rozvany, editor, *Optimization of Large Structural Systems*, volume 1, pages 455–476. Kluwer, 1993.

[4] M. Zyczkowski and A. Gajewski. Optimal structural design under stability constraints. In J. M. T. Thompson and G. W. Hunt, editors, *Collapse: The buckling of structures in theory and practice*, Cambridge, 1982. IUTAM, Cambridge University Press.

[5] A. Forsgren and U. Ringertz. On the use of a modified Newton method for nonlinear finite element analysis. *Computer Methods in Applied Mechanics and Engineering*, 110:275–283, 1993.

[6] U. T. Ringertz. On the design of Beck's column. *Structural Optimization*, 8:120–124, 1994.

[7] K. Isogai. Direct search method to aeroelastic tailoring of a composite wing under multiple constraints. *J. Aircraft*, 26:1076–1080, 1989.

[8] M. H. Shirk, T. H. Hertz, and T. A. Weisshaar. Aeroelastic tailoring - theory, practice, and promise. *J. Aircraft*, 23:6–18, 1986.

[9] M. L. Overton. Large-scale optimization of eigenvalues. *SIAM J. on Optimization*, 2:88–120, 1992.

[10] P. E. Gill, W. Murray, M. A. Saunders, and M. H. Wright. Two steplength algorithms for numerical optimization. Report SOL 79-25, Department of Operations Research, Stanford University, 1979.

[11] A. V. Fiacco and G. P. McCormick. *Nonlinear Programming*, volume 4 of *Classics in applied mathematics*. SIAM, Philadelphia, 1990.

[12] F. Alizadeh. *Combinatorial optimization with interior point methods and semi-*

definite matrices. PhD thesis, Department of Computer Science, University of Minnesota, 1991.

[13] F. Jarre. An interior-point method for minimizing the maximum eigenvalue of a linear combination of matrices. Report SOL 91-8, Department of Operations Research, Stanford University, 1991.

[14] U. T. Ringertz. Optimal design of nonlinear shell structures. Report FFA TN 91-18, The Aeronautical Research Institute of Sweden, 1991.

[15] U. T. Ringertz. An algorithm for optimization of nonlinear shell structures. *Int. J. Num. Meth. Eng.*, 38:299–314, 1995.

[16] U. T. Ringertz. Numerical methods for optimization of eigenvalues. In *Proceedings 5th AIAA/NASA/USAF/ISSMO Symposium on Multidisciplinary Analysis and Optimization*, Panama City, Florida, September 1994. AIAA 94-4359.

[17] S. Boyd, L. El Ghaoui, E. Feron, and V. Balakrishnan. *Linear matrix inequalities in system and control theory*, volume 15 of *Studies in applied mathematics.* SIAM, 1994.

[18] P. E. Gill, W. Murray, M. A. Saunders, and M. H. Wright. Maintaining LU-factors of a general sparse matrix. *Linear Algebra and its Applications*, 88/89:239–270, 1987.

[19] U. T. Ringertz. Optimization of eigenvalues in nonconservative systems. In G. I. N. Rozvany and N. Olhoff, editors, Proceedings of the first world congress on structural and multidisciplinar optimization. Pergamon, pp. 741–748, 1995.

[20] L. N. Trefethen, A. E. Trefethen, and S. C. Reddy. Pseudospectra of the linear Navier-Stokes evolution operator and instability of plane Poiseuille and Couette flows. TR 92-1291, Department of Computer Science, Cornell University, 1992.

[21] R. Byers. A bisection method for measuring the distance of a stable matrix to the unstable matrices. *SIAM J. on Scientific and Statistical Computing*, 9:875–881, 1988.

OPTIMIZATION ISSUES IN OCEAN ACOUSTICS

A. TOLSTOY*

Abstract. Ocean acoustics, e.g., SONAR, has traditionally been used for the detection and localization of targets such as submarines or schools of fish. However, more recently the community has focused on the use of acoustics to probe the ocean itself and its boundaries. Ocean acoustic tomography for the purpose of estimating temperature structure throughout a large ocean volume is a fine example of such an application. This a a case where theory, propagation modeling, technology, and computer capabilities have all reached a sufficient level of maturity that such a large scale inverse problem becomes tractable. Other applications include the estimation of shallow water bottom properties such as sediment thicknesses and densities, of under-ice reflectivity, and more generally the estimation of any parameters which influence the acoustic propagation. However, the community continues to struggle with such basic issues as how to pose each problem properly so as to guarantee uniqueness for the solution and how to find that optimizing solution. The essential difficulty in finding solutions arises not only because the search space for the unknowns can be extremely large, but also because that space is usually highly non-convex thereby preventing the use of simple gradient based searches. Methods in use to find solutions include simulated annealing, genetic algorithms, and tailored search algorithms based on examinations of the solution space itself.

Key words. ocean acoustics, shallow water, tomography, non-convex, inversion algorithms, sound-speed profile, ocean temperature, acoustic field, matched field processing, environmental parameters, linearization, normal modes, geoacoustics.

1. Introduction to ocean acoustics. Sound propagation is a very complicated process in an ocean environment. Bounded above by the air (or ice) and below by the seafloor, these upper and lower boundaries usually reflect significant energy back into the water channel resulting in a waveguide for acoustic energy. The behavior of energy in this waveguide is affected by many environmental properties such as the temperature profile of the water which can vary as a function of depth, latitude, longitude, and time. Higher water temperatures result in faster acoustic propagation speeds. Pressure as a function of depth is another a crucial influence where higher pressures result in faster propagation speeds, and seawater salinity is also a factor influencing ocean sound-speed, although its effect is far less than either temperature or pressure. Fluctuations in sound-speed values also occur as a result of vertical displacements due to internal waves traveling through the water volume. Additional environmental factors include bubbles and biological scatterers, e.g., fish bladders, as well as the nature of the boundaries, particularly if the surfaces are rough as might be the case under Arctic ice (which actually introduces an *elastic*, rough layer), or under large storm waves, or over a bottom covered by coral or manganese nodules. The seafloor can also display significant topography which can

* Integrated Performance Decisions, 4224 Waialae Avenue, Suite 5-260, Honolulu, HI 96816.
Many thanks to ONR for their continued support.

specularly deflect coherent acoustic energy out of the vertical plane between source and receiver. Moreover, the seafloor is also acoustically penetrable, and thus, its subsurface properties such as sound-speed profiles, densities, porosity, permeability, elasticity, and embedded scatterers can also affect propagation depending on the acoustic frequencies of interest and the bottom parameter values. Motivations to study the environmental properties of the ocean vary from interests in the long term behavior of large scale ocean temperatures (to address global warming issues) to improved techniques for locating valuable resources (such as manganese nodules on the seafloor). Some of the acoustic concepts are illustrated in Fig. 1 where c, c_1, c_2 denote the acoustic sound-speeds (the reciprocals of the indices of refraction). Let us examine some typical behavior of acoustic energy. In

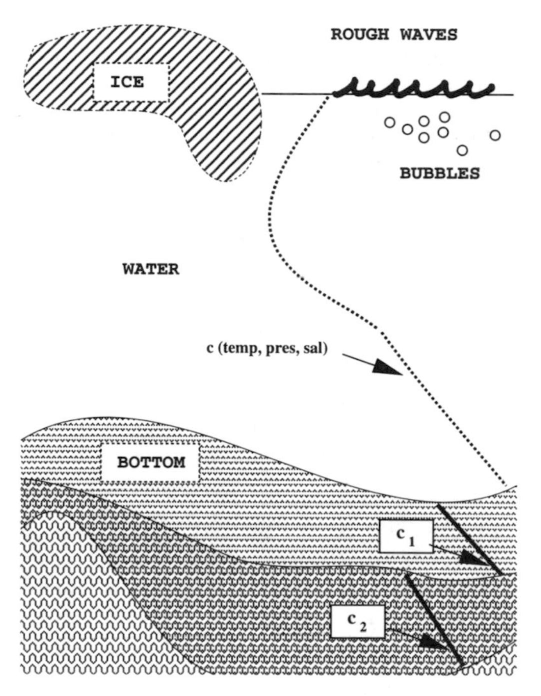

FIG. 1. *Illustration of the ocean environment and some of the important factors affecting acoustic propagation.*

Fig. 2 we see a characteristic deep ocean sound-speed profile $c(z)$ varying as a function of depth z. In particular, we see the sound-speed structure near the surface dominated by temperature which usually decreases with deeper depths until a depth is reached where temperature becomes constant. This depth is usually between 500 and 1000m. At this point pressure becomes the dominant influence increasing with depth. Sound-speeds in the ocean vary between 1450m/s and 1550m/s, i.e., less than 10%. However, even these seemingly small perturbations result in major effects, allowing for sound to refract away from the boundaries which would otherwise scatter and absorb the energy. Consequently, low frequency signals (10-100Hz) which correspond to the lowest order modes for the waveguide can propagate over tens of thousands of kilometers and still be sufficiently intense to be heard more than half a world away. In Fig. 3a we see a plot of acoustic intensity (assuming the sound-speed profile of Fig. 2) as a function of range and depth for a 20Hz point source located at the upper left of the plot at $z = 100$m depth, $r = 0$km range. The dark shades indicate high intensity values with the scale indicated in dB

$$1.0\text{dB} = 10\log(I(r,z)/I_o),$$

for $I(r,z)$ = intensity at (r,z), I_o = intensity at $r = 1$m). For this case we have assumed that the bottom is totally absorbing so that there are no reflections from it. We note how the energy appears to be cyclical with high intensity, well focused features appearing at intervals of about 50km. Moreover, there are also regions of very low intensity (known as shadow zones), particularly near the surface and bottom. This behavior is strictly the result of the sound-speed profile character shown in Fig. 2. If we modify the sound-speed profile to allow for something like a cold water eddy near the surface producing the sound-speed profile shown in Fig. 4, then the acoustic field changes significantly. In particular, in Fig. 3b we see the effects of this new profile indicating more acoustic energy trapped near the surface and producing noticeable changes in the entire intensity structure at all depths and ranges. As an illustration of the effects which the bottom might add, in Fig. 3c we see the field assuming all parameters are as for Fig. 3a *except* that the bottom allows for some reflectivity. One result is that the previous shadow zones are now filling in with energy and the interference patterns are becoming more intricate.

It is easy to conclude that since the oceans are so complex, it must be nearly impossible to describe acoustic propagation in such an environment with any confidence. However, there has been an amazing amount of success in acoustic prediction via numerical modeling over the last few decades, at least for deep water where the bottom effects are far less pronounced than for shallow water. There are now many models from which to choose (all based on the wave equation or the reduced wave equation) allowing for tradeoffs between computational efficiency and computational accuracy as relevant to the parameters important to the particular problem of interest (see [9]). For example, normal mode based codes are usually quite fast at low frequencies and very accurate for range *in*dependent environments while ray based codes are better suited for higher frequencies and in range dependent situations. The present day success in accurately predicting ocean acoustic fields has resulted in the successful application of these models to inverse problems for the determination of their *input* parameters.

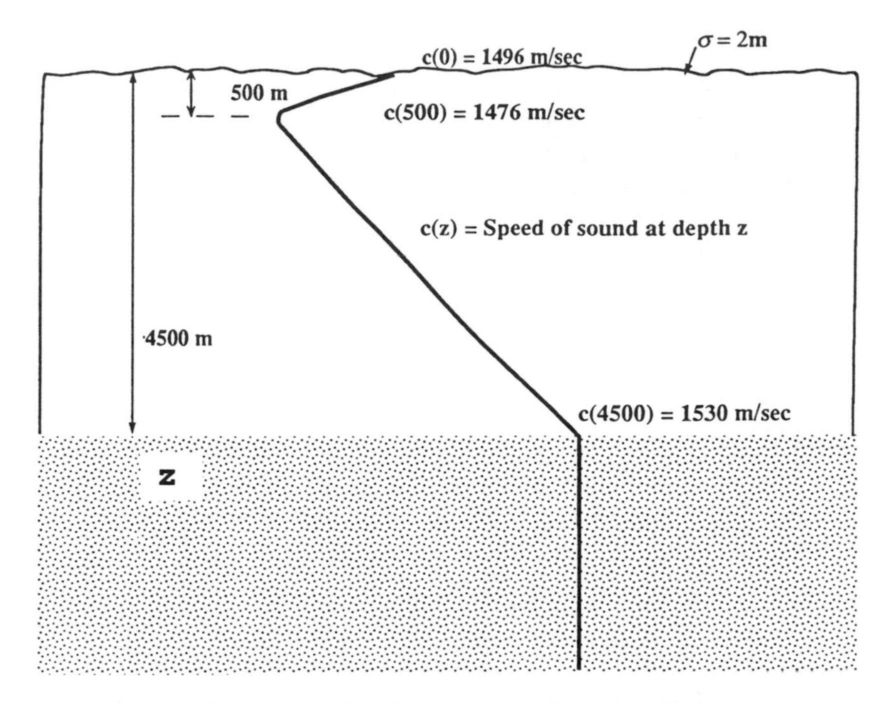

FIG. 2. *A typical deep ocean sound-speed profile.*

2. Important related inverse problems. For the past 50 years or so the study of ocean acoustic fields has been directed almost exclusively toward the detection and localization of targets. For such purposes, model inputs such as environmental parameters were the keys to successfully finding targets and were generally viewed as the weak links in the chain since it can be extremely difficult, time-consuming, and expensive to obtain such data. However, one person's stumbling block can be another's building block. In particular, within the last 15 years a new emphasis has developed in the Underwater Acoustics community where acoustic energy is now viewed as a wonderful remote sensing probe for the estimation of environmental parameters (at least for those which affect acoustic propagation) such as surface roughness and reflectivity, ocean temperatures, bottom sediment thicknesses, densities, and porosity.

Recent work which has used acoustics for the determination of such environmental parameters includes the estimation of:

- Surface roughness parameters [16].
- Arctic ice reflectivity [12].
- Ocean sound-speed profiles [4], [15], [21].
- Shallow water bottom properties [6]; using a simulated annealing based algorithm [3], [11], [5], [1]; using genetic algorithms [7], [8].

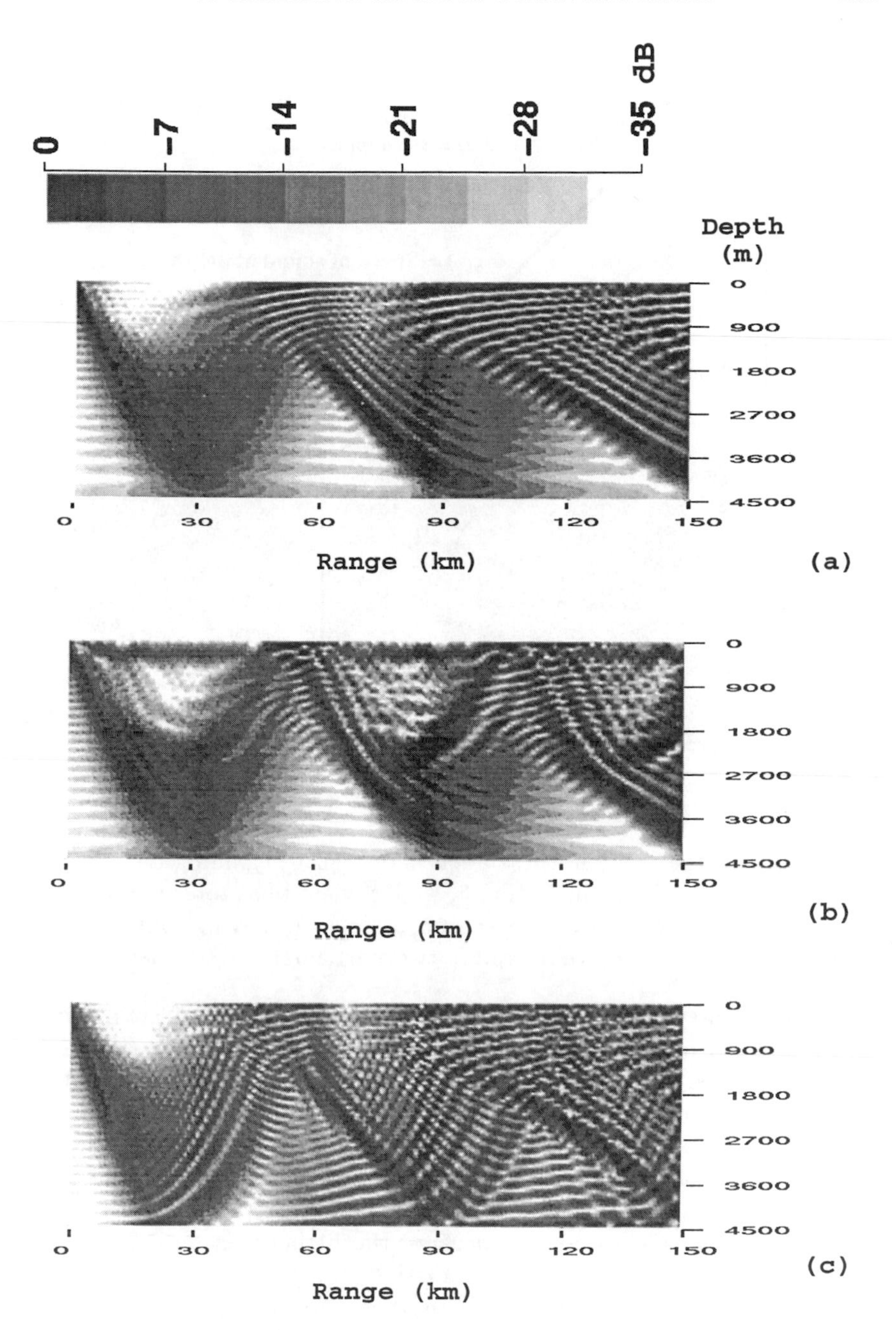

FIG. 3. *Plots of acoustic intensity as a function of range and depth for a 20Hz point source located at 0km range, 100m depth. For (a) and (b) the bottom is assumed to be totally absorbing. The scale is indicated in dB with dark shades corresponding to high intensity. (a) We assume the sound-speed profile shown in Fig. 2. (b) We assume the sound-speed profile shown in Fig. 4. (c) We assume the sound-speed profile of Fig. 2 plus some bottom reflectivity.*

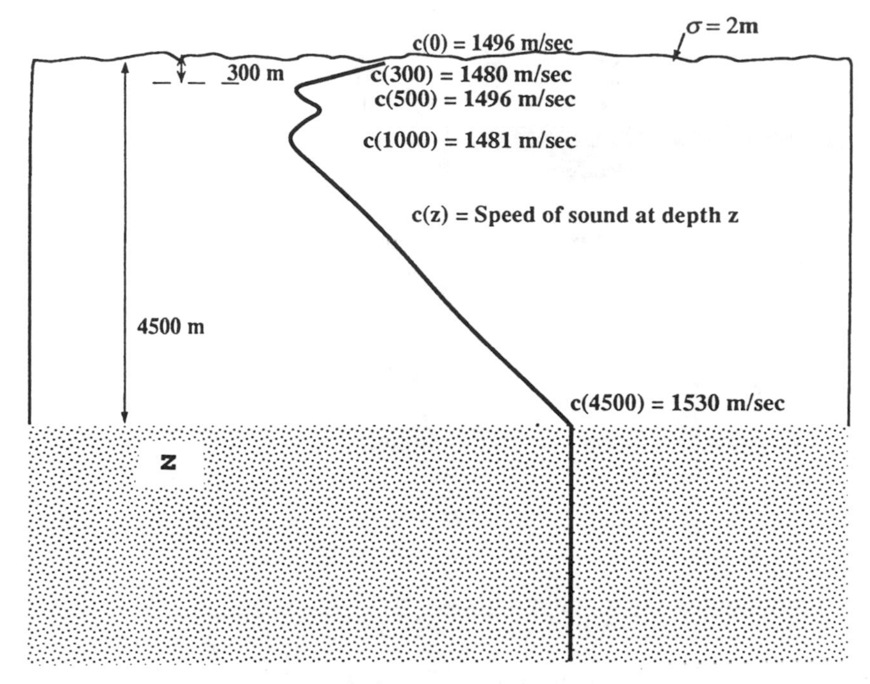

FIG. 4. *A deep ocean sound-speed profile showing a surface duct as might result from a cold eddy.*

It is fair to say that work in these areas is only just beginning but looks extremely promising. Next in section 3, we'll review what tools are presently available and being applied to ocean acoustic environmental inverse problems. Then in section 4, we'll explore two particular cases in detail.

3. Standard optimization tools. The estimation of environmental parameters involves studying the effects of many possible candidates for those values. If we begin with measured data from which to extract information about the parameters *plus* a trusted model to estimate that measured field as a function of the parameters, then we also need an inversion method to find those parameter values which "best fit" the data. Two fundamental tools used to build an inversion method include:

1. a way to quantitatively measure the fit between data and model (often referred to as a cost function),
2. a technique to search through the enormous space of candidate parameter values.

An important new way to compare data with model involves a signal processing technique known as Matched Field Processing, and important techniques to search through the solution space include: simulated annealing, genetic algorithms, designer algorithms which may involve directed searches, and a linearization of the problem to simplify the search process down to a matrix inversion. Details on these tools follow in subsections 3.1-3.4 next.

3.1. Matched field processing (MFP). The essence of MFP is to determine unknown parameters by comparing or correlating complex data, i.e., phases and amplitudes, received on an *array* to model predictions for such data where the unknown parameters of interest are varied as inputs to the model. Important reasons why MFP is such an attractive technique include:

- if a field is highly structured and non-planar, then this will *help* MFP performance given that the field can be accurately modeled;
- coherent array processing allows for improved *noise* suppression;
- only *relative* signal phases and amplitudes, i.e., receiver to receiver, need to be accurately modeled rather than absolute levels.

Consider as an example a known source at a known location generating an acoustic field which propagates to an array of receivers in a ocean with *unknown* parameters x_1, x_2. Then, these parameters are determined by MFP according to the following procedure:

1. collect *measurements* of the acoustic field on the array;
2. Fourier transform the data and select a single *low frequency*[1] having good signal-to-noise levels and good sensitivity to the parameters of interest;
3. compute *theoretical* acoustic fields such as would be heard on the array at that frequency for the known source using a propagation model with a family of input test values $\hat{x}_1, \hat{x}_2$ estimating the unknown parameters x_1, x_2;
4. cross-correlate the measured with the theoretically predicted fields. The values of $\hat{x}_1, \hat{x}_2$ producing the *highest* correlation should correspond to the "true" values x_1, x_2.

We can see this process illustrated in Fig. 5 (for a vertical array) with the cross-correlation output displayed graphically in the lower right as an "Ambiguity Surface" (AMS) where higher correlations are indicated by darker shades. The term "ambiguity" suggests that there are often multiple sets of parameter values producing highly similar acoustic fields as heard on the arrays thus resulting in high MFP correlation values. These ambiguities are immediately visible on an AMS, and when these areas of high correlation are not in the vicinity of the true values they are referred to as "sidelobes".

Describing the MFP technique as a simple correlation between data and model gives one an idea of the principle involved but is not strictly accurate. In most cases we examine more than one set of data points over time and subsequently consider the expected value of the cross-spectral

[1] Broadband or time domain data are sometimes treated directly but usually MFP is performed on a single frequency.

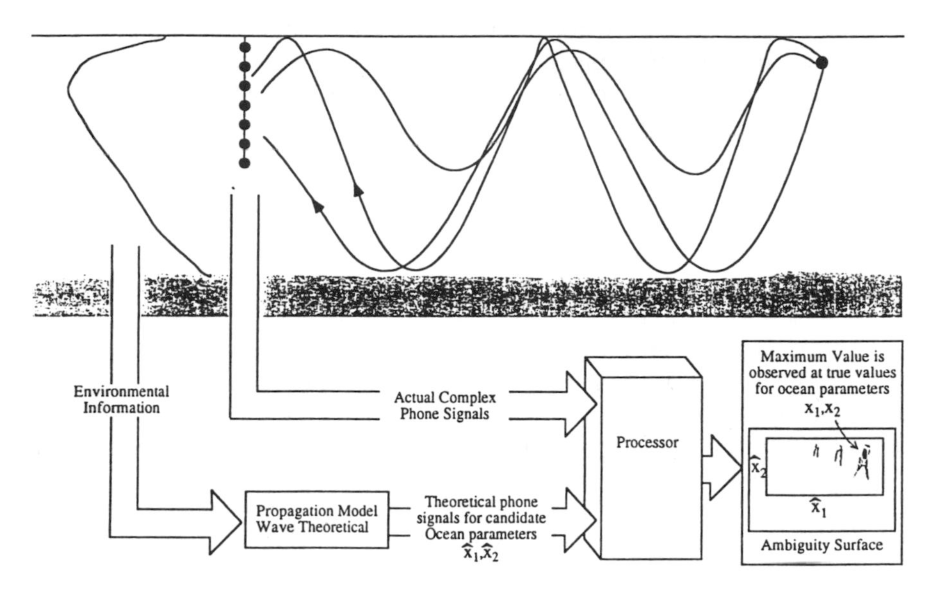

FIG. 5. *Illustration of the basic components of Matched Field Processing (MFP).*

matrix (resulting in the "Linear Processor") rather than a single snapshot value. In particular we have the output of the linear processor P_{lin} testing parameters $\hat{\mathbf{x}}$ given by

$$P_{lin}(\hat{\mathbf{x}}) = \hat{\mathbf{F}}^+ \langle \mathbf{F}\mathbf{F}^+ \rangle \hat{\mathbf{F}},$$

$$\approx |\hat{\mathbf{F}}^+ \cdot \mathbf{F}|^2,$$

where $\langle A \rangle$ is the expected value of A, $\mathbf{F} = (F_1, F_2, ..., F_n)^T$, F_i denotes the measured complex acoustic field at the ith receiver, $i = 1, 2, ..., n$, $\hat{\mathbf{F}} = (\hat{F}_1, \hat{F}_2, ..., \hat{F}_n)^T$, $\hat{F}_i$ denotes the estimated field at the ith receiver, and the fields have been normalized so that the L_2 norms are

$$\| \mathbf{F} \|, \| \hat{\mathbf{F}} \| = 1.$$

We note that maximizing $P_{lin}(\hat{x}) \approx |\hat{\mathbf{F}}^+ \cdot \mathbf{F}|^2$ is nearly equivalent to minimizing the least squares fit to data $1 - |\hat{\mathbf{F}}^+ \cdot \mathbf{F}|^2$.

Also, high resolution, non-linear processors are often considered in addition to the Linear Processor since they can offer increased sensitivity to the parameters of interest and improved suppression of sidelobes or false solutions. Moreover, such high resolution techniques do not rely on a least squares fit to data which has its own pitfalls. See [18] for complete details on MFP.

3.2. Simulated annealing. Simulated annealing (S.A.) is a well known optimization technique derived from an analogy with thermodynamics, i.e., the way that liquids can freeze into crystals of maximally ordered structures in minimum energy states [13], [14]. This approach is applied when the search space is non-convex, i.e., contains many local extrema which will stall a conventional gradient descent technique. The S.A. method allows a gradient descent solution to "bounce" out of its local minimum (rather like a hot molecule) and travel to another candidate minimum. The bounce of the solution is controlled by the user who can set a "temperature" which can be modified as the process progresses or "cools".

Key components of the simulation process involve:

- A function to be minimized/optimized, e.g., an MFP based cost function.
- A description of candidate system configurations, i.e., the solution space bounds and search steps.
- A random number generator used to select probabilistic options for the intermediate "solution" as it travels around the space.
- A control parameter/temperature used to set the likelihood that the "solution" will bounce out of a local minimum. The best versions allow for this parameter to change dynamically according to statistical information acquired during the process.

Some of the attractive features of the S.A. approach include:

- The method is not fooled by a quick payoff.
- It is simple computationally and easily programmed.
- It is nearly independent of the selected starting point.
- It finds families of "solutions".

While this method enjoys some popularity it has its disadvantages. In particular,

- There is no guarantee that "the" solution will be found.
- There is a significant amount of trial and error involved in tuning the controls, e.g., temperature, cooling schedule, parameter search increments (selection of neighbors).
- For any given problem the method may *not* be any better than an well designed progressive search.

3.3. Genetic algorithms. A somewhat newer optimization technique resembling S.A. has been incorporated into computational algorithms which are then known as genetic algorithms (G.A.s). The technique is based on the mechanics of natural selection combining a randomized search with an evolving population/family of candidates. Key components of such algorithms involve:

- A cost function to be minimized/optimized and which will determine the "fitness" of a candidate (and its subsequent probability of selection to reproduce).

- Selection of an initial population/family of candidates and their encoding into "genetic" strings.
- Operations, i.e., reproduction, crossover, and mutation, used to evolve the population into optimized or fitter candidates.
- A random number generator used to change the population:
  - Select candidates/pairs for reproduction.
  - Select portions of the "genetic strings" for each pair of candidates. Then, for crossover: switch the portions.
  - Select bits of "genetic strings" for infrequent random alteration (mutation).

G.A.s have the same advantages of S.A. discussed in the previous section. In addition, they

- search with a population of samples rather than a tracking a single point;
- have lots of intercommunication between individuals/samples;
- can provide statistics on the "*a posteriori* likelihood" of the candidates as optimal solutions;
- offer lots of flexibility for tuning.

On the down side, they have the same problems as S.A. plus:

- they require encoding problem into a "genetic type" format;
- they offer lots of flexibility for tuning.

3.4. Designer algorithms and linearization. Designer algorithms are tailor-made routines which take maximal advantage of *a priori* information about the solution space and the nature of the optimizing function. Such routines are presently used for the MFP inversion for shallow water bottom properties (which we will briefly discuss in section 4.2 next with inversion details in [20]) and were also used in the early stages of the MFP deep water tomography [21], [22]. One major drawback to such algorithms is that they have sudden, unexpected pitfalls which can cause abrupt stalling or other misbehaviors.

Some problems are amenable to reformatting in terms of a linear system of equations. This can be an ideal way to solve for unknowns and offers a wide menu of well tested algorithms to solve the subsequent system. The trick is to embed the problem into matrix form and not throw out any *essential* non-linearities. An example of this approach is the latest version of the MFP tomography for large, deep ocean volumes which we will describe in section 4.1 next. The details of the associated inversion algorithm can be found in [17]).

4. Examples. In the sections to follow we will discuss: (1) MFP based deep ocean tomography (which proposes to sense mesoscale oceanographic features by means of multiple vertical arrays and air-deployed shot sources) and (2) a particular approach still under development for the use of MFP in the estimation of shallow water bottom properties such as sediment layer thicknesses, sound-speeds, and densities. This latter method is being applied to experimental data with tantalizing success.

We note that in designing the MFP based algorithms for the selected inverse problems, care must be taken to introduce as many constraints and to parameterize the unknowns as efficiently as possible. The design of the search algorithm needs to eliminate invalid solutions or sidelobes rapidly while maximizing convergence and computational speed. The search spaces can be enormous, e.g., if six parameters are unknown and each can have any of 10 possible values, then the result is one million possible combinations where the acoustic field may need to be computed for *each* candidate. Moreover, the search spaces are notoriously non-convex. That is, simple gradient based algorithms will not converge to the global maximum solution but rather will iterate toward false, local maxima.

4.1. MFP deep ocean tomography. MFP based deep ocean tomography proposes to sense 3-D ocean sound-speeds by means of multiple vertical arrays and air-deployed shot sources as illustrated in Fig. 6. The method relies on low frequency signals (from 10 to 30Hz) from many distributed sources at *known* depths and ranges and transmitting throughout the ocean volume. The low frequencies have the advantage that they are relatively impervious to the effects of rough sea surface scattering and of internal waves. Additionally, they reduce the accuracy requirements placed on source and array localizations (locations need to be known within a wavelength λ, e.g., we have $\lambda = c/f = 75$m where $c = 1500$m/s, $f = 20$Hz.).

In addition to the operational/hardware requirements, there are several conceptual factors involved in the MFP approach where the final goal is to find the family of sound-speed profiles representing the actual fronts and eddies, i.e., the profiles which maximize the MFP power as heard on the arrays. This is the computationally difficult part of the problem, and it is made easier by:

1. efficiently characterizing the unknowns (via Empiricial Orthogonal Functions[2] (EOFs) described in [10]);
2. optimizing the array-source geometry (described in [17]);
3. linearizing the problem wherever possible (described in [17]).

The approach involves a number of steps after selecting an ocean region of interest. The first step is to grid the region into cells such that the unknown environmental parameters for each cell (such as a depth-dependent profile or a sediment layer thickness) can be assumed to be fixed throughout that cell. In the examples to date these cells have been assumed to be

[2] The EOF basis functions have the advantageous feature that they are the *most efficient* basis functions (in a least squares sense) for an expansion of the sound-speed profiles. In general, only 2 or 3 such functions are necessary for a highly accurate representation of a profile. Thus, our parameter search space has been reduced to finding 2 or 3 coefficients per sound-speed profile rather than finding all the values of the sound-speed profile as a function of depth. Of course, pathological profiles, i.e., those whose character differs significantly from the profiles generating the EOFs, may require more than 2 or 3 terms for their expansions.

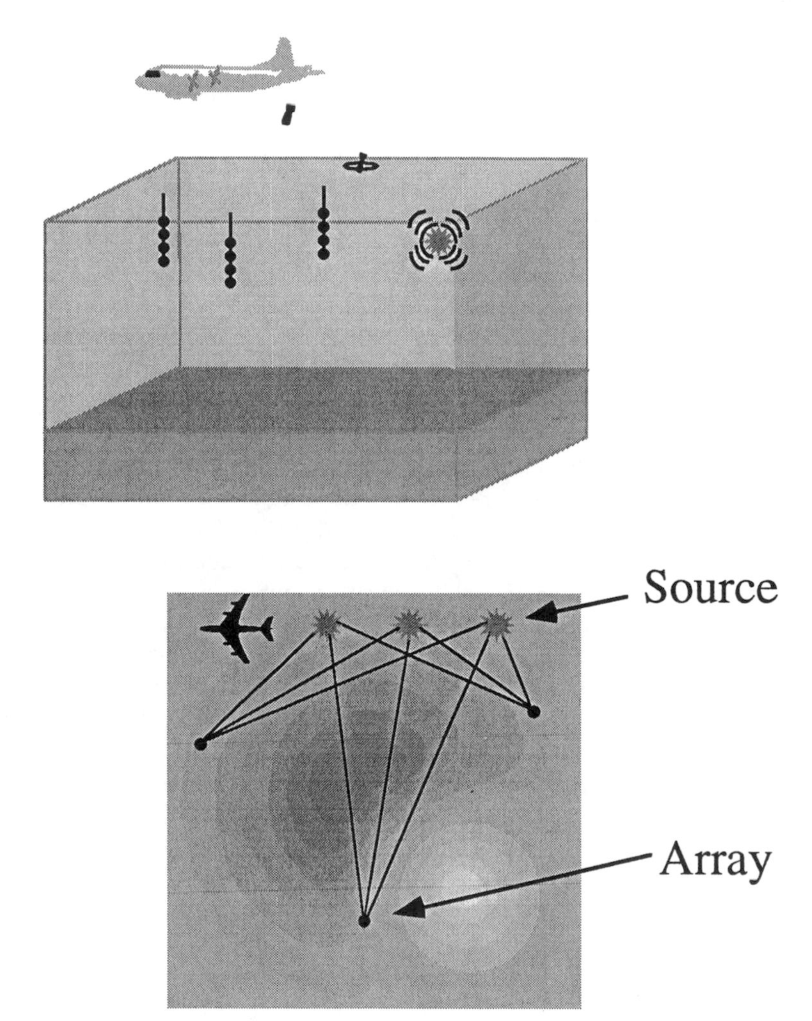

FIG. 6. *Illustration of MFP acoustic tomography method. (a) An airplane drops shot sources which sink and explode below the surface. The resulting acoustic signal is measured by several vertical arrays of receivers/hydrophones floating in the ocean. (b) An overhead view of the ocean region such as seen from the airplane and showing the acoustic transmission paths from the explosive sources to the vertical arrays. Also shown are two ocean eddies, indicated by the light and dark shadings.*

simple, uniform rectangular tubes (square in range and cross-range) but could be non-uniformly tailored to any given scenario in order to more finely sample areas of high variability.

Next, sources and arrays are distributed around and through the region in a configuration carefully designed by preliminary simulations to optimize the acoustic sampling of all the region cells. The size of the ocean region, the number of cells, the nature of the parameters being investigated, and the desired inversion accuracy all influence the number of sources and receivers and their distributions.

Finally, one must select an appropriate frequency, i.e., a frequency high enough to excite many modes and yield sufficient sound-speed resolution but low enough that attenuation and sensitivity to random and unknown perturbations are minimized. Then, in combination with an accurate but computationally rapid propagation model, e.g., adiabatic normal modes [9], the inversion algorithm operates by comparing modeled fields with the measured fields. In particular, the algorithm seeks the family of parameters representing the unknown environment which *simultaneously* maximizes the correlation between measured and modeled relative acoustic amplitudes and phases across the arrays for *all* of the source-receiver paths.

In 1992 C.S. Chiu [2] developed a simulated 3-D sound-speed data set for a deep ocean composed from four simulated oceanographic modes and representing a highly realistic, 1000km by 1000km by 5000m ocean region. In Fig. 7a we see a slice at depth 300m through one quarter of that ocean and showing the sound-speed structure. This data set shows considerable variability and was designed in order to help quantitatively evaluate/benchmark tomographic methods.

The MFP approach was applied to this data set. First, for each depth dependent profile we generated the highly efficient EOFs, and their associated range and cross-range dependent coefficients. We found that only two EOFs were needed to characterize each profile to accuracies of better than 0.15m/s. Thus, we only needed to determine two parameters for each cell. Next, since the full region is quite large, it was divided into quarters: four subsets 500km by 500km by 5000m (one of which we see in Fig. 8a). Then, we gridded each subregion into 400 cells each 25km by 25km by 5000m and "deployed" the sources and arrays as shown in Fig. 8 with arrays each 1050m long consisting of 15 phones at 75m intervals (spaced at half-wavelengths for a 20Hz signal) with the first phone near the surface. Next, using an adiabatic normal mode model we generated the "data", i.e., the acoustic phases and amplitudes on the arrays for each source given the exact/true environment. Finally, assuming that we knew the sound-speeds for the cells containing arrays and sources we generated inversion estimates of the more than 600 unknowns on a Silicon Graphics 2010A 36MHz Workstation using 24 hrs of CPU to generate the results shown in Fig. 7b. Comparing Figs. 7a and 7b we note that the inversion has captured a remarkable level of detail and mimics all significant features.

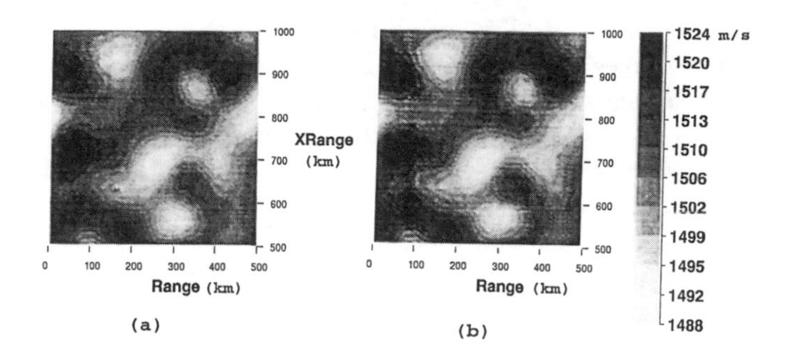

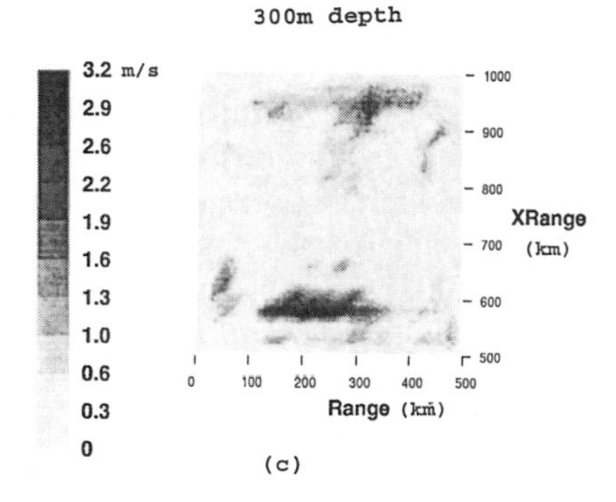

FIG. 7. *Plot of Chiu sound-speeds at 300m depth for one quarter of the full simulated ocean. a) Plot of true Chiu sound-speeds. Gray scale shows a maximum value of 1524m/s and a minimum value of 1488m/s. b) Plot of sound-speeds as estimated by the MFP inversion using the configuration shown in Fig. 8 with processing at 20 Hz. c) Error of MFP inversion. Gray scale shows a maximum of 3.2m/s.*

The quantitative differences are shown in Fig. 7c where we find that the maximum difference is 3.2m/s with an overall rms error level of 0.66m/s. In general, the inversion for this quadrant shows the largest errors of the four subregion inversions. A variety of simulated resources and likely test errors were subsequently considered and showed that the inversion is stable with respect to diminished resources and expected errors. In general, fewer resources and/or larger errors for the *apriori* information result in degraded estimates.

4.2. MFP shallow water geoacoustic inversion. The most recent application of the MFP approach for environmental inversion has been for the estimation of shallow water bottom properties. While the approach is similar to that used for the estimation of deep ocean sound-speed profiles, more parameters need to be determined per cell, the frequencies to be analyzed need to be higher, the source to array ranges must be much

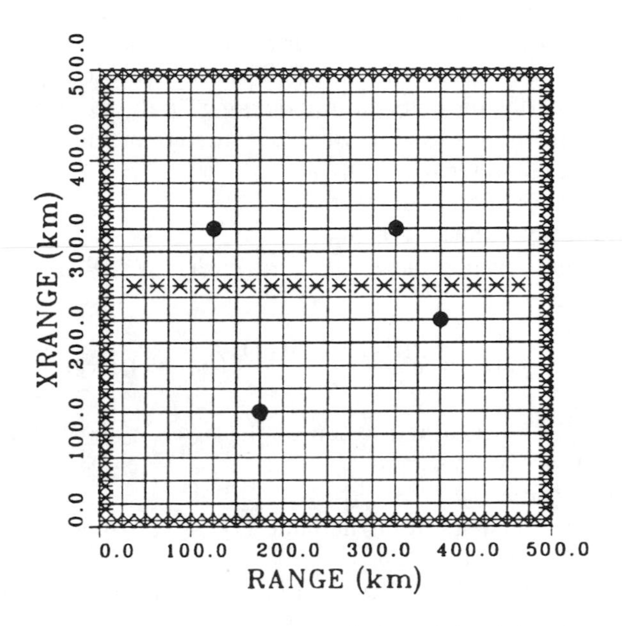

FIG. 8. *Optimal distribution of four arrays (indicated by o's) and 174 shot sources (indicated by ⋆'s) for 400 cell region.*

shorter, and the field sensitivities to the individual parameters will vary by scenario.

While the long term goal of this work is to determine bottom properties in a fully 3-D environment where all parameters including bottom depth can vary, that goal is a long way from being achieved. A major obstacle is the fact that variable topography will result in horizontal refraction/out-of-plane propagation which can be of major significance in such situations. Not only is there a scarcity of propagation models which can *accurately* predict the amplitude and phase of a signal in such an environment, but those codes which are available are so computationally demanding as to rule out their use in the MFP inversion which can require tens of thousands of replications. However, some MFP simulation work [19] has been done to date for shallow water environments which allow for 3-D variability of bottom properties – except for topography. This rather artificial problem allows us to investigate the ability of the method to converge to a smaller subset of parameters than might be encountered in a real test situation. We can also use a fast propagation code such as adiabatic normal modes whose use would not be appropriate in the presence of topographic features.

In this paper we will discuss a very simple scenario: no range or azimuthal variability, a single source, and a single vertical array. Work is in progress to estimate such bottom parameters using actual experimental data for this kind of scenario. However, a full discussion of that effort is beyond the scope of this paper (see [20]), so we shall confine ourselves here primarily to a few basic issues involving simulated data.

In general, the determination of shallow water bottom properties can involve the estimation of many parameters, e.g., a dozen or more, for each range independent path segment. Such parameters include:

- the average water depth h_{water},
- the number of bottom layers N,
- the nth layer thicknesses h_n,
- the sound-speed profiles $c_n(z)$ for each layer where we assume a linear form

$$c_n(z) = \frac{(c_n(z_1) - c_n(z_2))}{h_n}(h_n - z) + c_n(z_2)),$$

with z_1 the depth at the top of the layer, z_2 is the depth at the bottom of the layer,
- layer densities ρ_n,
- the final sound-speed of the underlaying half-space c_{bot},
- the half-space density ρ_{bot}.

This leads to a total of $M = 4(N + 1)$ bottom parameters for each range independent segment. If each of these parameters could only have 10 possible values each, we would have a search space consisting of 10^M candidates. For two layers of sediment this results in 10^{12} candidates. In addition, there are also parameters involving the source and array positions plus the water sound-speed profile (since they are often not known with sufficient accuracy at the higher frequencies required for shallow water situations). Hence, it is critical to optimize the search algorithm. We begin our inversion efforts by examining acoustic sensitivity AMSs as functions of the parameters to be estimated. From these AMSs we determine not only which are the key parameters but also how finely to sample the search space for each parameter. Subsequently, we can select the range and increments of variability considered to be "reasonable" for our scenario. Finally, examining sensitivity AMSs we can get an understanding of how the acoustic fields vary throughout the search space as a function of these parameters, and this information can be critical in converging to the optimizing parameter values. The full inversion process for shallow water bottom properties is *not yet* automated, but rather remains a combination of hands-on experience and focussed computational searches.

As an example consider a shallow water environment shown in Fig. 9 with water depth h_{water} of 115.5m over a bottom consisting of two sediment layers, the first of thickness $h_1 = 2.6$m with a linear sound-speed profile varying from $c_1(115.5) = 1620$m/s at the top of the sediment layer to $c_1(118.1) = 1662$m/s at the bottom with constant density $\rho_1 = 1.15$, the second sediment layer of known thickness 100m with a linear sound-speed profile varying from $c_2(118.1) = 1601$m/s at the top to $c_2(218.1) = 1691$m/s at the bottom with constant density $\rho_2 = 1.35$, and which overlays a non-elastic half space with a constant sound-speed $c_{bot} = 1691$m/s and density $\rho_{bot} = 1.35$. Let us also assume an array of 9 phones spaced every 8m (producing an array 64m long) with the top phone at 35.72m depth (bottom phone at 99.72m). Then, consider a point source at a range of 12.1km, depth of 20m generating a 100Hz frequency component (for which there are only 5 or 6 modes). In Fig. 10 we see AMSs showing the associated sensitivity of the inversion problem (for this source and receiver array) to the indicated pairs of parameters where the non-varying parameters are fixed. Computations have also been constrained to only consider combinations for which $c_n(z_1) \leq c_n(z_2)$ and for which $\rho_1 \leq \rho_2$. For example, in Fig. 10a we see the sensitivity to the sound-speed parameters $c_1(115.5), c_1(118.1)$ of the thin sediment where the other parameter values are fixed at their correct values. In Fig. 10 we note that the parameters which are most likely to be correctly determined are $h_1, c_2(118.1)$ while insensitivity to $c_1(z)$ will result in only the average sediment sound-speed of the first layer being determined. We note also that the surfaces are non-convex (this is most clearly seen in Fig. 10b).

Based on much preliminary investigation of sensitivity AMSs actual experimental data has been inverted resulting in very promising results as indicated in Fig. 11 where the MFP response level improved from the 0.05 value shown in Fig. 11a (assuming nominal values for the environmental parameters as suggested by previous surveys) to a value to 0.68 as seen in Fig. 11b (assuming the values suggested by the inversion optimization) out of a maximum possible value of 1.00. Moreover, the source position was also significantly improved and located near its nominal position of 12.1km with error bars as indicated.

5. Conclusions and future work. Acoustics as a remote sensing tool for the study of oceans is a relatively new field. It is only recently that acoustic propagation models have become sufficiently accurate and computer processors have become sufficiently powerful to allow consideration of acoustic environmental inversion problems. However, we are finding that there is still much room for improvement with regard to models for shallow water. Moreover, the parameter search spaces can be enormous and seriously non-convex. Thus, in addition to efforts to improve the quality and speed of the acoustic models the next critical improvements will hopefully be in the realm of search and convergence techniques for finding of the "optimizing" solutions.

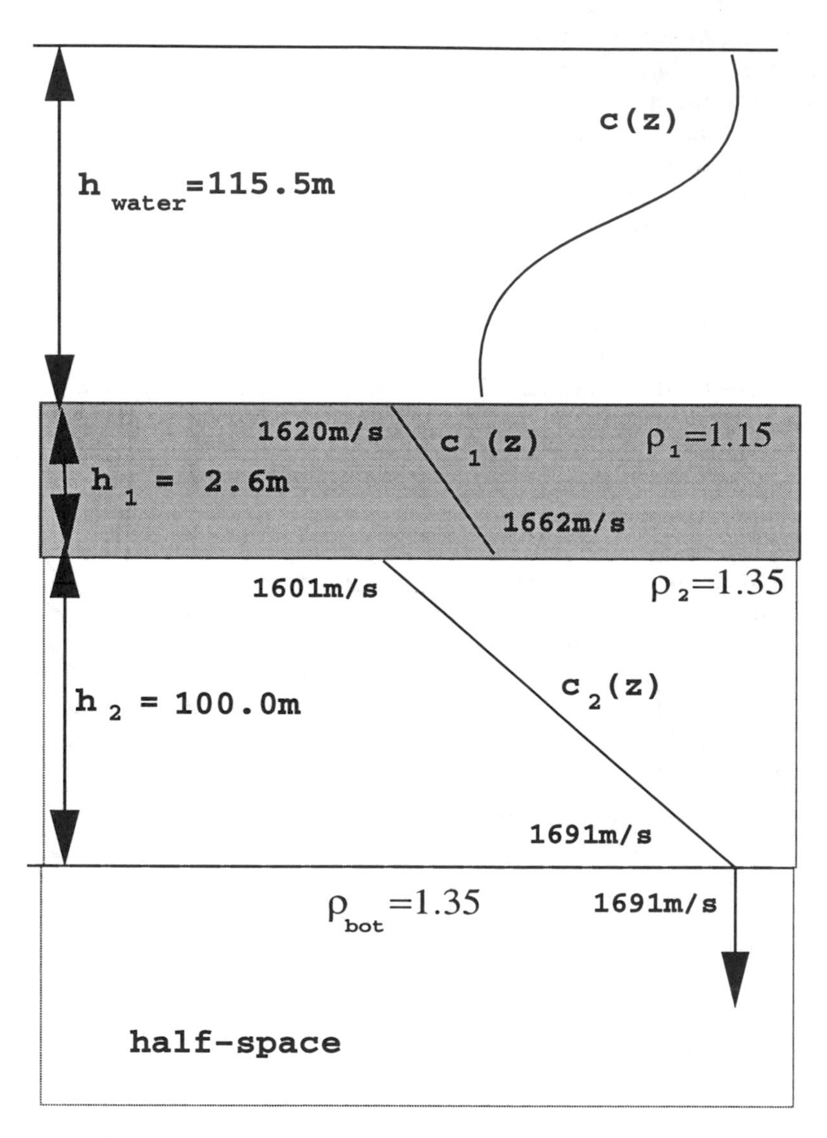

FIG. 9. *Shallow water parameters as used for simulated data.*

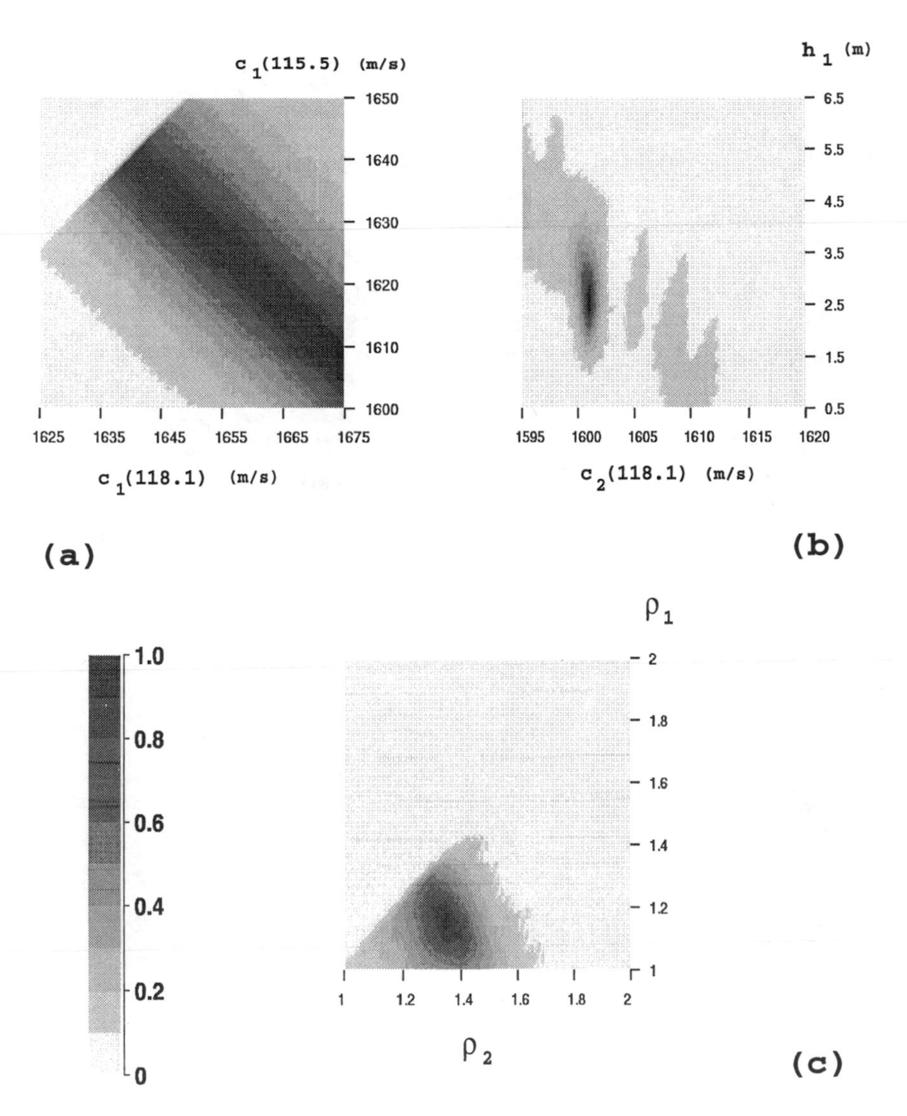

FIG. 10. *Shade plots of high resolution Ambiguity Surfaces (AMSs) showing acoustic sensitivity to bottom parameters at 100Hz. MFP response scale is indicated with black at ideal maximum value = 1.00, white at minimum = 0.00. All non-varying parameters are fixed at their* true *values. (a) Sediment layer 1 sound-speeds* $c_1(115.5)$ *versus* $c_1(118.1)$. *MFP peak = 1.00 at* $c_1(115.5) = 1622m/s$, $c_1(118.1) = 1660m/s$. *(b) Sediment layer 1 thickness* h_1 *versus sediment layer 2 sound-speed* $c_2(118.1)$. *MFP peak = 1.00 at* $h_1 = 2.6m$, $c_2(118.1) = 1601m/s$. *(c) Sediment layer 1 density* ρ_1 *versus sediment layer 2 density* ρ_2. *MFP peak = 1.00 at* $\rho_1 = 1.15, \rho_2 = 1.35$.

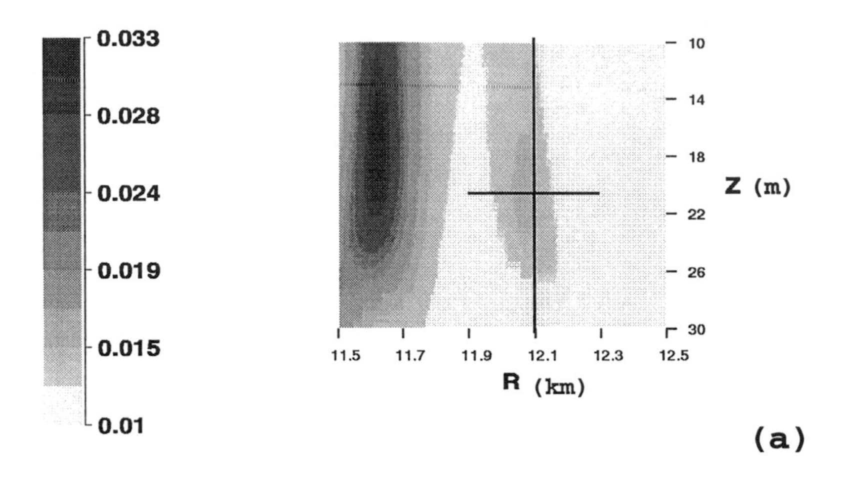

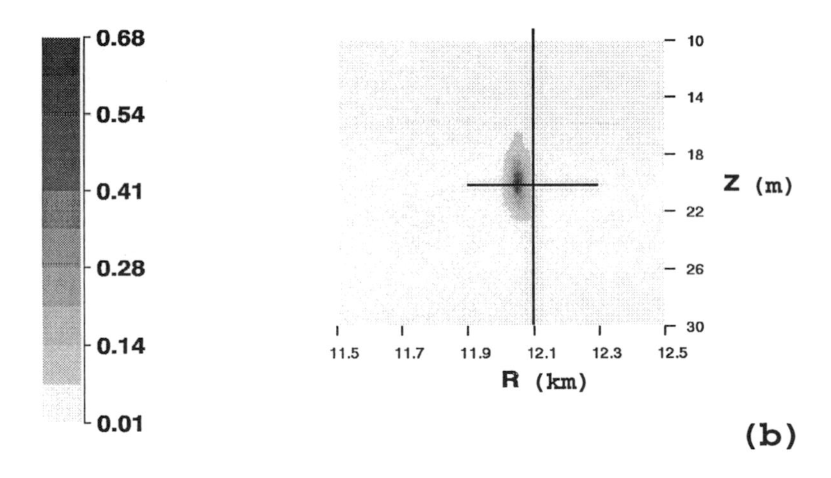

FIG. 11. *AMSs showing improved MFP peak values and source localization for experimental data. (a) The nominal environmental parameter values were used. (b) The improved parameter values suggested the inversion optimization were used.*

REFERENCES

[1] CHAPMAN, N.R. AND K.S. OZARD, *Matched field inversion for geoacoustic properties of young crust*, in *Full Field Inversion Methods in Ocean and Seismic Acoustics*, ed. O. Diachok et al., Kluwer Academic Pubs, 1995.

[2] CHIU, C.S., *Acoustical Oceanography: Comparison of Ocean Tomography Methods*, Volume 92, J. Acoust. Soc. Am., 1992, pp. 2324.

[3] COLLINS, M.D., W.A. KUPERMAN, AND H. SCHMIDT, *Nonlinear inversion for ocean-bottom properties*, volume 92, J. Acoust. Soc. Am., 1992, pp. 2770–2783.

[4] CORNUELLE, B., W. MUNK, AND P. WORCESTER, *Ocean acoustic tomography from ships*, volume 94, J. Geophys. Res., 1989, pp. 6232–6250.

[5] DOSSO, S.E., M.L. YEREMY, J.M. OZARD, AND N.R. CHAPMAN, *Estimation of ocean-bottom properties by matched field inversion of acoustic field data*, volume 18, IEEE J. Oceanic Eng., 1993, pp. 232–239.

[6] FRISK, G.V., *Inverse methods in ocean bottom acoustics*, in *Les Houches Session L Oceanographic and Geophysical Tomography*, North-Holland, 1990.

[7] GERSTOFT, P., *Global inversion by genetic algorithms for both source position and environmental parameters*, volume 3, J. Computat. Acoust., 1994, pp. 251–266.

[8] GINGRAS, D.F. AND P. GERSTOFT, *Inversion for geometric and geoacoustic parameters in shallow water: Experimental results*, volume 97, J. Acoust. Soc. Am., 1995, pp. 3589–3598.

[9] JENSEN, F.B., W.A. KUPERMAN, M.B. PORTER, AND H. SCHMIDT, *Computational Ocean Acoustics*, Am. Inst. Physics, 1994.

[10] LEBLANC, L.R. AND F.M. MIDDLETON, *An underwater acoustic sound velocity data model*, volume 67, J. Acoust. Soc. Am., 1980, pp. 2055–2062.

[11] LINDSAY, C.E. AND N.R. CHAPMAN, *Matched field inversion for geoacoustic model parameters using adaptive simulated annealing*, volume 18, IEEE J. Oceanic Eng., 1993, pp. 224–231.

[12] LIVINGSTON, E. AND O. DIACHOK, *Estimation of average under-ice reflection amplitudes and phases using matched field processing*, volume 86, J. Acoust. Soc. Am., 1989, pp. 1909–1919.

[13] METROPOLIS, N., A.W. ROSENBLUTH, M.N. ROSENBLUTH, A.H. TELLER, AND E. TELLER, *Equation of state calculations by fast computing machines*, volume 21, J. Chem. Phys., 1953, pp. 1087–1092.

[14] PRESS, W.H., B.P. FLANNERY, S.A. TEUKOLSKY, AND W.T. VETTERING, *Numerical Recipes*, Cambridge University Press, 1986.

[15] SHANG, E.C., *Ocean acoustic tomography based on adiabatic mode theory*, volume 85, J. Acoust. Soc. Am., 1989, pp. 1531–1537.

[16] TOLSTOY, A., *Matched field estimation of environmental parameters*, 21st Annual Congress of Canadian Meteorological and Oceanographic Soc., 1987.

[17] TOLSTOY, A., *Linearization of the matched field processing approach to acoustic tomography*, volume 91, J. Acoustic. Soc. Am., 1992, pp. 781–787.

[18] TOLSTOY, A., *Matched Field Processing for Underwater Acoustics*, World Scientific Pub, 1993.

[19] TOLSTOY, A., *Matched field tomographic inversion to determine environmental properties*, in *Current Topics in Acoustical Research*, ed. Council of Scientific Info., Research Trends, 1994, pp. 53–61.

[20] TOLSTOY, A., *Using matched field processing to determine shallow water environmental parameters*, in *Theoretical and Computational Acoustics '95*, Lee et al. ed., 1996, pp. 283–295.

[21] TOLSTOY, A., O. DIACHOK, AND L.N. FRAZER, *Acoustic tomography via matched field processing*, volume 89, J. Acoustic. Am., 1991, pp. 1119–1127.

[22] TOLSTOY, A. AND L.N. FRAZER, *Maximizing matched field processing output used in a new approach to ocean acoustic tomography*, published in *Computational Acoustics*, Vol 2, Lee, Robinson, and Vichnevetsky, ed., North-Holland, 1993, pp. 79–84.

GRADIENT METHODS IN INVERSE ACOUSTIC AND ELECTROMAGNETIC SCATTERING

P.M. VAN DEN BERG* AND R.E. KLEINMAN†

Abstract. The problem of determining the complex permittivity or sound speed in a bounded inhomogeneity imbedded in a homogeneous medium from scattered field measurements exterior to the inhomogeneity is considered. A number of methods of attacking this problem, all based on minimizing the difference between an integral representation of the scattered field and the measured data, are described. These include Born, Newton-Kantorovich and distorted Born methods. The main part of the paper will be devoted to a description of a gradient type algorithm which is used to minimize a cost functional in which two objective functions are sought simultaneously. The error which is minimized is a bilinear form involving the product of two functions. This special form of the nonlinearity is retained in the algorithm. A number of numerical results will be presented which illustrate the effectiveness and limitations of the approach.

1. Introduction and formulation. A large class of inverse problems concerns the determination of constitutive material parameters within a bounded object immersed in a homogeneous medium from measurements of the field scattered when the object is illuminated by a known single frequency wavefield. The unknown parameter is usually the index of refraction which may be complex if the medium is lossy. The starting point in the class of problems discussed here is the domain integral equation, sometimes called the Lippmann–Schwinger equation, which governs the wave process within the object.

This integral equation is used in conjunction with an integral representation of the field exterior to the object to express the discrepancy between the measured field and the predicted field as a functional of the desired material parameters. Finding these parameters to minimize this functional is the object of the various methods treated here.

We consider the two-dimensional case wherein the object, B, is assumed to be cylindrical of arbitrary bounded cross section. Let D denote the interior of a bounded domain in $\mathbb{R}^2$ with piecewise smooth (e.g. piecewise C^2) boundary ∂D (with normal $\boldsymbol{\nu}$). A Cartesian coordinate system is centered in D. Denote points in $\mathbb{R}^2$ as $\boldsymbol{p} = (x_p, y_p)$ and $\boldsymbol{q} = (x_q, y_q)$. We assume that the unknown scatterer, B, is contained in the domain D. This constitutes a priori information about the approximate location of the scattering object. D may be chosen quite large to insure inclusion of B but this incurs a computational price.

We assume that the object is irradiated successively by a number of known incident fields $u_j^{\text{inc}}(\boldsymbol{p})$, $j = 1, 2, \cdots, J$. For each incident field the

---

* Laboratory of Electromagnetic Research, Faculty of Electrical Engineering, Delft University of Technology, P.O. Box 5031, 2600 GA Delft, The Netherlands.

† Center for the Mathematics of Waves, Department of Mathematical Sciences, University of Delaware, Newark, DE 19716, U.S.A..

This work was supported under AFOSR Grant No. F49620-94-1-0219.

total field will be denoted by $u_j(\boldsymbol{p})$ in D and by $u_j^{\text{ext}}(\boldsymbol{p}) = u_j^{\text{inc}}(\boldsymbol{p}) + u^{\text{sct}}(\boldsymbol{p})$ exterior to D. The wave fields are governed by the differential equations

$$[\nabla^2 + k^2 n^2(\boldsymbol{p})]\, u_j(\boldsymbol{p}) = 0 \quad \text{in} \quad D\,, \tag{1.1}$$

$$[\nabla^2 + k^2]\, u_j^{\text{sct}}(\boldsymbol{p}) = 0 \quad \text{ext}\, D\,, \tag{1.2}$$

the transmission conditions

$$u_j^{\text{ext}}(\boldsymbol{p}) = u_j(\boldsymbol{p}) \;\text{ and }\; \frac{\partial}{\partial\nu}\, u_j^{\text{ext}}(\boldsymbol{p}) = \frac{\partial}{\partial\nu}\, u_j(\boldsymbol{p}) \quad \text{on } \partial D\,, \tag{1.3}$$

and the radiation condition

$$\frac{\partial}{\partial r}\, u_j^{\text{sct}} - iku_j^{\text{sct}} = o\left(r^{-\frac{1}{2}}\right) \quad \text{as} \quad r := |\boldsymbol{p}| \to \infty\,, \tag{1.4}$$

where k denotes the constant wave number of the exterior medium. More precise restrictions are given in [2], Here u_j denotes either the excess acoustic pressure or the non vanishing component (z–component) of the electric field. The index of refraction, n, is given by

$$n^2 = \frac{c^2}{c_D^2(\boldsymbol{p})} \;\text{(acoustics), or }\; n^2 = \frac{\epsilon_D(\boldsymbol{p})}{\epsilon} \;\text{(TM electromagnetics)}, \tag{1.5}$$

where c and c_D are wave speeds and ϵ and ϵ_D are permittivities exterior and interior to D, respectively. We assume the relative permeability of the two media is one. It has been shown (e.g. [7]) that fields satisfying equations (1.1)–(1.4) also satisfy the following integral relations

$$u_j(\boldsymbol{p}) = u_j^{\text{inc}}(\boldsymbol{p}) + \frac{ik^2}{4}\int_D \chi(\boldsymbol{q})\, u_j(\boldsymbol{q})\, H_0^{(1)}(k|\boldsymbol{p}-\boldsymbol{q}|)\, dv(\boldsymbol{q}),\ \boldsymbol{p} \in \text{int}D, \tag{1.6}$$

$$u_j^{\text{sct}}(\boldsymbol{p}) = \frac{ik^2}{4}\int_D \chi(\boldsymbol{q})\, u_j(\boldsymbol{q})\, H_0^{(1)}(k|\boldsymbol{p}-\boldsymbol{q}|)\, dv(\boldsymbol{q}),\ \boldsymbol{p} \in \text{ext}D\,, \tag{1.7}$$

where the contrast $\chi(\boldsymbol{q})$ is defined as

$$\chi(\boldsymbol{q}) := n^2(\boldsymbol{q}) - 1\,, \tag{1.8}$$

and clearly $\chi(\boldsymbol{q}) = 0$ at those points in D exterior to the actual scattering object B.

In the inverse scattering problem u_j^{sct} will be measured on some surface S which includes D in its interior, so equation (1.7) is written symbolically as the data equation,

$$u_j^{\text{sct}} - G_S\, \chi u_j = 0\,,\ \ \boldsymbol{p} \in S\,, \tag{1.9}$$

where the operator G_S is defined implicitly by (1.7). Similarly the integral equation (1.6) implicitly defines the operator G_D so that (1.6) may be written as the object equation,

$$u_j^{\text{inc}} - u_j + G_D \, \chi u_j = 0 \,, \quad \boldsymbol{p} \in D \,. \tag{1.10}$$

The subscripts S and D are appended to the integral operator G to clarify where in space the field point $\boldsymbol{p}$ is located.

In TE electromagnetics (magnetic field polarized along the cylinder axis), equation (1.1) is replaced by

$$\nabla \cdot \left(\frac{1}{n^2} \nabla u_j \right) + k^2 u_j = 0 \text{ in } D \,, \tag{1.11}$$

the transmission conditions (1.3) become

$$u_j^{\text{ext}}(\boldsymbol{p}) = u_j(\boldsymbol{p}) \text{ and } \frac{\partial}{\partial \nu} u_j^{\text{ext}}(\boldsymbol{p}) = \frac{1}{n^2} \frac{\partial}{\partial \nu} u_j(\boldsymbol{p}) \text{ on } \partial D \,, \tag{1.12}$$

and in place of (1.6) and (1.7) we have

$$u_j(\boldsymbol{p}) = u_j^{\text{inc}}(\boldsymbol{p}) + \frac{i}{4} \int_D \mathcal{M}(\boldsymbol{q}) \, \nabla_q H_0^{(1)}(k|\boldsymbol{p} - \boldsymbol{q}|) \cdot \nabla_q u_j(\boldsymbol{q}) \, dv(\boldsymbol{q}) \,, \tag{1.13}$$

when $\boldsymbol{p} \in D$, while for $\boldsymbol{p} \in S$ we have

$$u^{\text{sct}}(\boldsymbol{p}) = \frac{i}{4} \int_D \mathcal{M}(\boldsymbol{q}) \nabla_q H_0^{(1)}(k|\boldsymbol{p} - \boldsymbol{q}|) \cdot \nabla_q u_j(\boldsymbol{q}) \, dv(\boldsymbol{q}) \,, \tag{1.14}$$

which can be written symbolically as

$$u_j^{\text{inc}} - u_j + \tilde{G}_D \, \mathcal{M} u_j = 0 \,, \quad \boldsymbol{p} \in D \,, \tag{1.15}$$

$$u_j^{\text{sct}} - \tilde{G}_S \, \mathcal{M} u_j = 0 \,, \quad \boldsymbol{p} \in S \,, \tag{1.16}$$

with the operators $\tilde{G}_S$ and $\tilde{G}_D$ defined according to (1.13) and (1.14) as integral-differential rather than pure integral operators and

$$\mathcal{M}(\boldsymbol{p}) = 1 - \frac{\epsilon}{\epsilon_D(\boldsymbol{p})} = 1 - \frac{1}{n^2(\boldsymbol{p})} = \frac{\chi(\boldsymbol{p})}{1 + \chi(\boldsymbol{p})} \,. \tag{1.17}$$

As before it is assumed that permittivity can vary within B but permeability is constant throughout $\mathbb{R}^2$.

If the scattered field is measured on S to be $f_j(\boldsymbol{p})$, which includes measurement error, noise and any other signal contamination, then equation (1.9) will not in general be satisfied if f_j replaces u_j^{sct}. In fact we use this data equation to define the discrepancy between the measured data and the predicted scattered field corresponding to χ and u_j in D. Thus we define the data error to be

$$\rho_j(\boldsymbol{p}) := f_j(\boldsymbol{p}) - (G_S\,\chi u_j)(\boldsymbol{p}), \quad \boldsymbol{p} \in S. \tag{1.18}$$

If u_j and χ do not solve the integral equation (1.10) (the object equation) then we define an object error by

$$r_j(\boldsymbol{p}) := u_j^{\text{inc}}(\boldsymbol{p}) - u_j(\boldsymbol{p}) + (G_D\,\chi u_j)(\boldsymbol{p}), \quad \boldsymbol{p} \in D. \tag{1.19}$$

The inverse scattering problem consists of determining $\chi(\boldsymbol{p})$ from a knowledge of the incident fields, $u_j^{\text{inc}}(\boldsymbol{p})$, and $f_j(\boldsymbol{p})$ on S. In general this problem is both nonlinear and highly ill-posed, failing to fulfill any of Hadamard's conditions for well-posedness. Stability is compromised since the map from the field in $L^2(D)$ to the field on $L^2(S)$ is compact, hence the inverse map from the data to the field in D is unbounded. Actually, the field on S is C^∞, hence any data noise in $L^2(S) \setminus C^\infty(S)$ will ensure non-existence. Uniqueness is in some ways a more interesting question. If the values $f_j(\boldsymbol{p})$ are exact scattering data for an object then there clearly is a solution of the inverse problem. Whether there is only one solution is a subject of ongoing research. In $\mathbb{R}^3$ it is known that for plane wave incidence, $u^{\text{inc}} = \exp(ik\boldsymbol{p}\cdot\hat{\boldsymbol{\alpha}})$, if u^{sct} is measured exactly on S for every direction $\hat{\boldsymbol{\alpha}}$ then there is only one $\chi(\boldsymbol{p})$ which will give rise to these u^{sct}, that is, there is a unique solution of the inverse problem [7]. No results are available if u_j^{sct} is known on S for only a finite number of incident directions $\hat{\boldsymbol{\alpha}}_j$. Moreover in $\mathbb{R}^2$ the uniqueness property has not been established even if u^{sct} is known for all incident directions.

Despite the lack of rigorous uniqueness results, the inverse problem has been cast as an optimization problem with some considerable success. We denote the norm and inner product in $L^2(D)$ and $L^2(S)$ by appending a subscript D or S as appropriate, i.e., $\|\cdot\|_D$, $\langle\cdot,\cdot\rangle_D$, etc. Then the inverse scattering problem is recast as an optimization problem of finding $\chi \in U_{ad}$ to minimize $\|\rho_j\|_S$, $j = 1, 2, \cdots, J$, subject to the constraint that the object equation (1.10) is satisfied in some sense. The space of admissible functions, U_{ad}, should incorporate any a priori information and in the absence of any it is taken as $L^\infty(D)$. The existence of a minimizer can be guaranteed by suitable choice of U_{ad}, however uniqueness remains an ongoing concern.

In TE electromagnetics a data error and a corresponding object error may be defined using (1.15) and (1.16). Thus corresponding to (1.18) and (1.19) we have

$$\tilde{\rho}_j(\boldsymbol{p}) = \tilde{f}_j(\boldsymbol{p}) - (\tilde{G}_S\mathcal{M}\,u_j)(\boldsymbol{p}), \; p \in S, \tag{1.20}$$

and

$$\tilde{r}_j(\boldsymbol{p}) = \tilde{u}_j^{\text{inc}}(\boldsymbol{p}) - (\tilde{G}_D\,\mathcal{M}\,u_j)(\boldsymbol{p}), \; p \in D. \tag{1.21}$$

Although there has been some work on the TE case in electromagnetics e.g., [16,26], much more has been done in the zero frequency limit where the inverse problem is known as Electrical Impedance Tomography and treated by methods different than those discussed here (e.g., [4,21,27]). When $k \neq 0$ much more attention has been focussed on the TM case which, while simpler in that the integral operators on D do not also involve differentiating the field, still retain the essential nonlinearity and ill-posedness of the inverse problem. Some of the methods employed are described in the next sections. Wherever possible, the methods will be described in terms of continuous models but it should be kept in mind that all numerical implementation requires a discretization, or projection onto a finite-dimensional space. If the domain D and the surface S are partitioned into N and M subdomains respectively and, in the simplest implementation, u_j, χ and $\mathcal{M}$ are taken to be piecewise constant on each subdivision of D, the discrete form (1.9, 1.10) and (1.15, 1.16) involve known vectors u_j^{inc} of length N, $u_j^{\text{sct}} = f_j$ of length M and unknown vectors u_j, χ and $\mathcal{M}$ of length N, and matrices G_D, $N \times N$, and G_S, $M \times N$. Because the operator G_D is of convolution type, the associated matrix is Toeplitz and the matrix vector operation is efficiently performed using a discrete FFT routine.

The first group of methods all involve substituting the formal solution of the object equation (1.10),

$$u_j = (I - G_D\chi)^{-1}\, u_j^{\text{inc}}\,, \tag{1.22}$$

into the data equation (1.9), and the data error (1.18) so that the data error becomes

$$\rho_j = f_j - G_S\chi(I - G_D\chi)^{-1}\, u_j^{\text{inc}}\,. \tag{1.23}$$

2. Born approximation. When the scatterer is small or the index of refraction is close to one (so called weak scatterers) then the well-known Born approximation consists of taking the field, u_j in D, to be the incident field. This in fact is the first term in a Neumann series solution of the object equation, that is, approximating $(I - G_D\chi)^{-1}$ in (1.23) by I. With this approximation the data error becomes

$$\rho_j^B = f_j - G_S\, \chi u_j^{\text{inc}} \quad \text{on} \quad S\,, \tag{2.1}$$

and the inverse problem is posed as determining χ as

$$\chi^B = \begin{matrix} \arg\min \\ \chi \in U_{ad} \end{matrix} \; \frac{\sum_{j=1}^{J} \|\rho_j^B\|_S^2}{\sum_{j=1}^{J} \|f_j\|_S^2}\,. \tag{2.2}$$

Other normalizations of the cost functional are possible. This is a classic example of an ill-posed problem and is treated by adding Tykhonov regularizers to the cost functional. The effectiveness of this approach is severely limited by the approximation of the field in D.

3. Iterative Born. Iterative improvement of the Born approximation is achieved with the following approach [23]. A sequence of functions $\{\chi_n\}$ is determined through the following algorithm

$$\chi_0 = 0, \quad \chi_{n+1} = \underset{\chi \in U_{ad}}{\arg\min} \frac{\sum_{j=1}^{J} \|f_j - G_S\,\chi(I - G_D\,\chi_n)^{-1}\,u_j^{\mathrm{inc}}\|_S^2}{\sum_{j=1}^{J} \|f_j\|_S^2},$$

or

$$\chi_{n+1} = \underset{\chi \in U_{ad}}{\arg\min} \frac{\sum_{j=1}^{J} \|f_j - G_S\chi\,u_{j,n}\|_S^2}{\sum_{j=1}^{J} \|f_j\|_S^2}, \tag{3.1}$$

where $u_{j,n}$ is a solution of

$$(I - G_D\,\chi_n)\,u_{j,n} = u_j^{\mathrm{inc}}. \tag{3.2}$$

As in the Born approximation, the optimization problem is solved by adding regularizers to the cost functional. It should be pointed out that in this approach, calculation of the quantity $(I - G_D\,\chi_n)^{-1}\,u_j^{\mathrm{inc}}$ is required and this means that the object equation, (3.2), must be solved at each iteration. This is a feature of a number of iterative methods which can be computationally challenging without a very efficient forward solver. Even with a good solver the work at each step will be increased and this will be more noticeable if a large number of iterations are used. Methods which avoid the need for a direct solution at each iteration will be described subsequently.

Initial guesses other than $\chi_0 = 0$ may be used. This is true of any of the iterative methods described here. Because of the lack of uniqueness, especially in the finite-dimensional versions of the optimization algorithms, one expects the occurrence of local minima in the cost functional. Improving the initial guess by incorporating as much a priori information as possible is often looked on as a way of avoiding being trapped in such local minima. However this does not diminish the importance of building in a priori information into the cost functional itself whenever possible.

4. Newton-Kantorovich-distorted Born. The cost functional to be minimized in the iterative Born method may be obtained by neglecting all but the zero-order term in an expansion of

$$(I - G_D\chi)^{-1} = [I - G_D(\chi_n + \delta\chi)]^{-1} \tag{4.1}$$

in powers of $\delta\chi = \chi - \chi_n$. If terms of order $O(\delta\chi)$ are retained in an expansion of equation (1.23) we obtain

$$\rho_j = f_j - G_S\chi_n u_{j,n} - G_S[I + \chi_n(I - G_D\chi_n)^{-1}G_D]\delta\chi\,u_{j,n} + O(\delta\chi^2). \tag{4.2}$$

Since it may be shown that

$$I + \chi_n(I - G_D\chi_n)^{-1}G_D = (I - \chi_n G_D)^{-1}, \tag{4.3}$$

the equation for ρ_j may be rewritten as

$$\rho_j = f_j - G_S\chi_n u_{j,n} - G_S(I-\chi_n G_D)^{-1}\delta\chi\, u_{j,n} + O(\delta\chi^2)\,. \tag{4.4}$$

Defining

$$\rho_{j,n} = f_j - G_S\chi_n u_{j,n} \tag{4.5}$$

leads to the so-called Newton-Kantorovich algorithm, used by Roger [17], Tabbara et al. [22] and others,

$$\chi_0 = 0, \quad \chi_{n+1} = \chi_n + \delta\chi\,, \tag{4.6}$$

with

$$\delta\chi = \begin{array}{c}\arg\min\\ \delta\chi \in U_{ad}\end{array} \frac{\sum_{j=1}^{J} \|\rho_{j,n} - G_S(I-\chi_n G_D)^{-1}\delta\chi\, u_{j,n}\|_S^2}{\sum_{j=1}^{J} \|f_j\|_S^2}\,, \tag{4.7}$$

and $u_{j,n}$ is again determined as the solution of the direct problem, (3.2). As before the optimization problem is solved by adding regularizers to the cost functional.

A scheme equivalent to the Newton-Kantorovich scheme is the distorted Born algorithm (e.g., [6]), which may be stated as follows. If χ_n is known, define an approximate Green's function as the solution of

$$(I - G_D\chi_n)G_n(\boldsymbol{p},\boldsymbol{q}) = \frac{ik^2}{4} H_0^{(1)}(k|\boldsymbol{p}-\boldsymbol{q}|)\,. \tag{4.8}$$

Let us further define the integral operator $G_{S,n}$ as

$$G_{S,n}f = \int_D G_n(\boldsymbol{p},\boldsymbol{q})\, f(\boldsymbol{q})\, dv(\boldsymbol{q})\,, \quad \boldsymbol{p} \in S\,. \tag{4.9}$$

From the governing differential equations we arrive at two alternative integral representations of the nth approximation of the scattered field, viz.,

$$u_{j,n} - u_j^{\text{inc}} = G_{S,n}\chi_n u_j^{\text{inc}}\,, \tag{4.10}$$

$$u_{j,n} - u_j^{\text{inc}} = G_S\chi_n u_{j,n}\,. \tag{4.11}$$

Now, in the distorted Born algorithm, the contrast is updated in the form $\chi_{n+1} = \chi_n + \delta\chi$, while

$$\delta\chi = \begin{array}{c}\arg\min\\ \delta\chi \in U_{ad}\end{array} \frac{\sum_{j=1}^{J} \|\rho_{j,n} - G_{S,n}\delta\chi\, u_{j,n}\|_S^2}{\sum_{j=1}^{J} \|f_j\|_S^2}\,, \tag{4.12}$$

where $\rho_{j,n}$ is defined in (4.5). The equivalence of this algorithm with the Newton-Kantorovich method given in (4.7) is established by showing that

$$G_S(I - \chi_n G_D)^{-1} = G_{S,n}\,, \tag{4.13}$$

which is not obvious. The discrete version of this identity is established in [11].

5. System formulation. All of the methods presented thus far involved utilizing various approximations to the formal solution of the object equation, (1.22), in the data error (1.23). The optimization problem then involved finding updates to the contrast by minimizing the norm of the data error. Conceptually this is attractive since it apparently reduces the size of the optimization problem. However, this advantage is offset by the fact that the field is either held fixed (the Born approximation) which severely limits the range of contrasts which may be reconstructed, or the field is updated at each iteration by solving a forward problem (1.22) with the most recent update of the contrast. The fact that large contrasts increase the nonlinear dependence of the fields and the data on the incident fields constitutes a serious limitation on all of the inversion algorithms described here unless a priori knowledge that the contrast is large is built into the inversion algorithm, an example of which is described below in Sect. 7.

An alternative approach involves the simultaneous solution of the system of equations

$$u_j - G_D \chi u_j = u_j^{\text{inc}}, \tag{5.1}$$

$$G_S \chi u_j = f_j, \qquad j = 1, 2, \cdots, J, \tag{5.2}$$

for the $J+1$ unknown functions χ and u_j in D. One attack on this problem was developed by Sabbagh and Lautzenheiser [19]. It consists of forming one unknown vector $(\chi, u_1, u_2, \cdots, u_J)$, whose dimension may be extremely large when the functions χ and u_j are discretized and solving the nonlinear equations by a standard nonlinear optimization algorithm. A few gradient based algorithms are described in [5]. This approach has the advantage of avoiding the necessity of solving the direct problems to update the fields, u_j, at each iteration. It has the disadvantage that the gradients are somewhat complicated and the matrix dimensions involved are large. One way to reduce the complexity is to treat the fields and the contrast separately as described in the next sections.

6. Modified gradient method. The basic idea of the modified gradient approach [12] for solving the inverse problem is the iterative construction of sequences $\{u_{j,n}\}$ and $\{\chi_n\}$, which converge to minimizers of the functional

$$F(\mathbf{u}, \chi) = w_D \textstyle\sum_{j=1}^{J} \|r_j\|_D^2 + w_S \textstyle\sum_{j=1}^{J} \|\rho_j\|_S^2, \tag{6.1}$$

where

$$w_D = \left(\textstyle\sum_{j=1}^{J} \|u_j^{inc}\|_D^2\right)^{-1} \quad \text{and} \quad w_S = \left(\textstyle\sum_{j=1}^{J} \|f_j\|_S^2\right)^{-1}, \tag{6.2}$$

and $\mathbf{u} = (u_1, u_2, \cdots, u_J)$. The weights w_D and w_S were chosen to balance the two error functionals which make up F_n in the sense that they are both equal to one if $u_j = \chi = 0$. This choice also ensures that the functional is insensitive to changes in the magnitude of the incident field and the data.

Specifically we define

$$(6.3)\qquad u_{j,n} = u_{j,n-1} + \alpha_n v_{j,n}\,, \quad \chi_n = \chi_{n-1} + \beta_n d_n\,, \quad n = 1, 2, \cdots .$$

For each n, the functions $v_{j,n}$ and d_n are update directions for the functions $u_{j,n}$ and χ_n, respectively, while the constants α_n and β_n are weights to be determined. We denote by $r_{j,n}$ and $\rho_{j,n}$ the residual errors obtained when $u_{j,n}$ and χ_n replace u_j and χ in (1.18) and (1.19). Further, F_n denotes the value of the cost functional F at the n-th iteration. The iterative algorithm will be completely specified when the starting values $u_{j,0}$ and χ_0, the update directions $v_{j,n}$ and d_n, and the weights α_n and β_n are specified. We employ the refined starting values and update directions described in [13]. These starting values are a "best" constant contrast and the associated fields, which are found by running the algorithm with $d_n = 1$ and $v_{j,n}$ either $r_{j,n-1}$ or the Polak-Ribière conjugate gradient direction (for fixed χ) according to a specified switching criterion. Once these starting values have been found, the update directions for the contrast are changed to the Polak-Ribière conjugate direction (for fixed u_j). The Polak-Ribière directions are given by

$$(6.4)\qquad v_{j,n} = g^v_{j,n} + \gamma^v_n v_{j,n-1}\,, \qquad \gamma^v_n = \frac{\sum_{j=1}^J \langle g^v_{j,n}\,,\, g^v_{j,n} - g^v_{j,n-1}\rangle_D}{\sum_{j=1}^J \|g^v_{j,n-1}\|^2_D}\,,$$

with the field gradient $g^v_{j,n} = \frac{\partial}{\partial u_j} F_{n-1}$ (assuming fixed contrast), i.e.,

$$(6.5)\qquad g^v_{j,n} = -2w_D(r_{j,n-1} - \overline{\chi}_{n-1} G^\star_D r_{j,n-1}) - 2w_S \overline{\chi}_{n-1} G^\star_S \rho_{j,n-1}\,,$$

and

$$(6.6)\qquad d_n = g^d_n + \gamma^d_n d_{n-1}\,, \qquad \gamma^d_n = \frac{\langle g^d_n\,,\, g^d_n - g^d_{n-1}\rangle_D}{\|g^d_{n-1}\|^2_D}\,,$$

with the contrast gradient $g^d_n = \frac{\partial}{\partial \chi} F_{n-1}$ (assuming fixed field), i.e.,

$$(6.7)\qquad g^d_n = 2w_D \textstyle\sum_{j=1}^J \overline{u}_{j,n-1} G^\star_D r_{j,n-1} - 2w_S \sum_{j=1}^J \overline{u}_{j,n-1} G^\star_S \rho_{j,n-1}\,,$$

where $G^\star_D$ and $G^\star_S$ are the adjoint operators mapping from $L^2(D)$ and $L^2(S)$, respectively, into $L^2(D)$ and the overbar denotes complex conjugate. Observe that the gradients are given in terms of the adjoint operators (Hermitean transposed matrices) which are known explicitly. Because the operator $G^\star_D$ is of convolution type, the associated matrix is Toeplitz and the matrix vector operation is efficiently performed using a discrete FFT routine. With the starting values and update directions now specified, the weights α_n and β_n are determined to minimize F_n. This nonlinear optimization problem is solved using a standard conjugate gradient method.

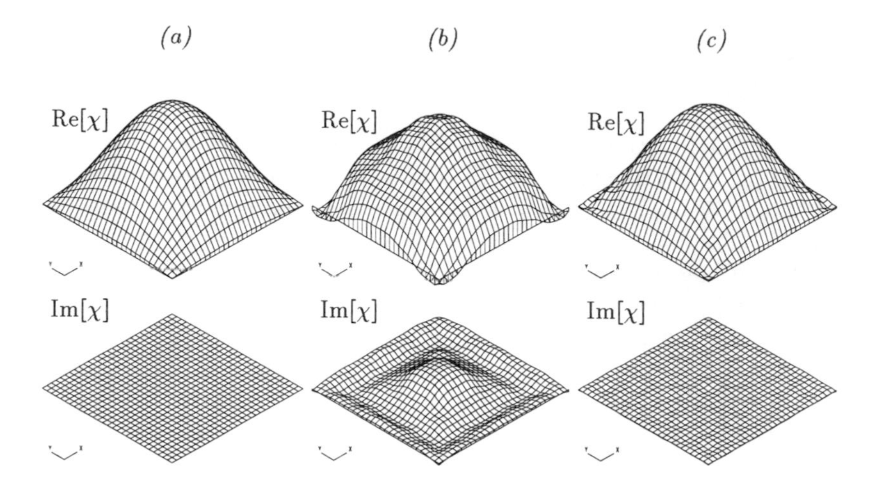

FIG. 6.1. *(a) The sinusoidal profile and the reconstructed profiles after (b) 64 iterations and (c) after 128 iterations.*

Note that the optimization problem is still ill-posed, nevertheless the presence of the object error in the cost functional seems to act as a regularizer to some degree as evidenced by the numerical performance of the algorithm presented in this and the following two sections. Additional regularizers are described in Sect. 9.

We illustrate the modified gradient method with a numerical example. The actual contrast was $\chi = \sin(\frac{\pi x}{3\lambda})\sin(\frac{\pi y}{3\lambda})$, $0 < x, y < 3\lambda$, see Fig. 6.1a. Synthetic "measured" data were found by numerical solution of the forward problem as described in [13]. The measurement surface S was chosen to be a circle of radius 9λ containing the test domain D which was a square, $d = 3\lambda$ by $d = 3\lambda$, where λ is the wavelength in the background (exterior) medium. Twenty nine stations ($J = 29$) were located uniformly on the surface S with each station serving successively as a line source and all stations acting as receivers. In the numerical examples the test domain was discretized into 29×29 subsquares. The reconstructions after 64 and 128 iterations are shown in Figs. 6.1b – 6.1c. Note that in the result after 64 iterations the imaginary part of the contrast has negative values, which is non-physical. For objects with larger contrasts ($kd|\chi_{\max}| \geq 6\pi$) this phenomenon is more severe and it helps to destroy the convergence of the inversion scheme. Without further a priori information, the present scheme successfully reconstructs complex contrasts up to $kd\chi_{\max} \approx 6\pi$. This upper limit may be extended dramatically by incorporating into the algorithm the fact that the imaginary part of the contrast is non-negative.

7. Location and shape reconstruction. With the inclusion of the non-negativity of the imaginary part of the contrast, the modified gradient method may also be used for reconstructing the location and shape of the boundary of an impenetrable object without making the a priori assumption of impenetrability [14]. The algorithm is precisely the same as that used for reconstructing the conductivity of a penetrable object and uses the fact that for high conductivity the skin depth of the scatterer is small, in which case the only meaningful information produced by the algorithm is the boundary of the scatterer. A striking increase in efficiency is achieved by incorporating into the algorithm the fact that for large conductivity, the contrast is dominated by a large positive imaginary part. This fact together with the knowledge that the scatterer is constrained in some test domain constitute the only a priori information about the scatterer that is used. There are no other implicit assumptions about the location, connectivity, convexity or boundary conditions. The non-negativity of the imaginary part of the contrast is incorporated into the algorithm by making the following assumptions:

$$\mathrm{Re}[\chi] = 0\,, \quad \text{and} \quad \mathrm{Im}[\chi] = \zeta^2\,, \quad \text{real } \zeta\,. \tag{7.1}$$

Instead of updating the contrast, $\chi_n = \chi_{n-1} + \beta_n d_n$, we update ζ as

$$\zeta_n = \zeta_{n-1} + \beta_n \xi_n\,, \tag{7.2}$$

with the directions ξ_n given by

$$\xi_n = g_n^\xi + \gamma_n^\xi \xi_{n-1}\,, \qquad \gamma_n^\xi = \frac{\langle g_n^\xi\,, g_n^\xi - g_{n-1}^\xi \rangle_D}{\|g_{n-1}^\xi\|_D^2}\,, \tag{7.3}$$

where g_n^ξ is the gradient of F with respect to changes in in ξ, evaluated at the $(n-1)$-st step, i.e.,

$$g_n^\xi = -2\zeta_{n-1}\mathrm{Im}[g_n^d]\,, \tag{7.4}$$

and g_n^d is given by (6.7). Note that the contrast gradient, $g_n^\xi(\boldsymbol{q})$, vanishes for zero values of $\zeta_{n-1}(\boldsymbol{q})$. We therefore cannot start the iterative scheme with a zero estimate for ζ_0. Furthermore, at locations in the domain where ζ_{n-1} vanishes, there is no new gradient direction. This latter fact is responsible for the success of the reconstruction of the shape of an object.

We illustrate the present method with a numerical example of the scattering by an off-centered circular cylinder. The measurement surface S is chosen to be a circle containing the test domain. We assume that the radius of this circle is large enough so that the far-field approximation of (1.18) may be employed, and the far-field coefficient is the quantity of interest so that the dependence on the radius is removed. We measure the far-field at 30 stations equally spaced around the object. Each of the stations serves

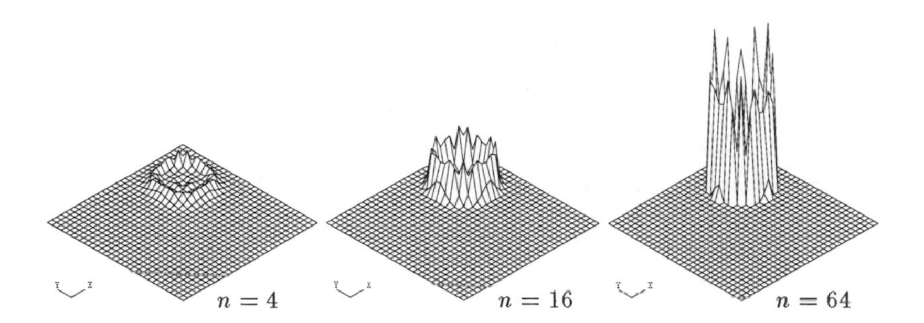

FIG. 7.1. *The reconstructed imaginary values of the contrast of the circular cylinder. At $n = 64$ the largest value is 8.9.*

in turn as the location of a source ($J = 30$) and the incident fields can be approximated as plane waves. The measured data were simulated by solving the direct scattering problem for an impenetrable circular cylinder. The analytic solution in terms of Bessel functions has been employed. The radius, a, of this circular cylinder was 0.015 m. In the example the test square was divided into 31×31 subsquares of 0.003×0.003 m^2. The wavelength is $\lambda = 0.030$ m, so that $ka = \pi$. Some surface plots of the reconstructed profiles (the imaginary part of the contrast χ) are presented in Fig. 7.1. After only four iterations the boundary of the object is clearly visible. Specifically we observe that after about 8 iterations the imaginary part of the contrast at the boundary becomes larger than 1 and only the contrast at the boundary of the object remains increasing when we increase the number of iterations. After 64 iterations the imaginary contrast at the boundary has reached values from 2.0 up to 8.9.

7.1. Bounded contrast reconstruction. In the preceding example we have seen that the modified gradient scheme indeed reconstructs the location and the shape of an impenetrable object by reconstructing the imaginary contrast at the boundary. However, the reconstructed contrast at the boundary becomes highly oscillatory after a couple of iterations. The peaks appear to increase with the number of iterations and it becomes difficult to choose the level value of the contour that estimates the boundary of the object. We therefore adopt a slightly modified reconstruction scheme. First of all we have observed that there is no improvement in locating the boundary after the contrast has reached a value such that the penetration depth of the wavefield is of the order of the mesh width in the testing domain. The visualization of the boundary of the object is improved when we impose an upper bound, $\chi_{\max}$, to the reconstructed contrast in such a way that the penetration depth of the wavefield is not less than three times the mesh width. This factor is chosen to provide a reconstructed object consisting of a 'boundary wall' with a thickness of two or three times the mesh width. If at some point in the iteration the reconstructed

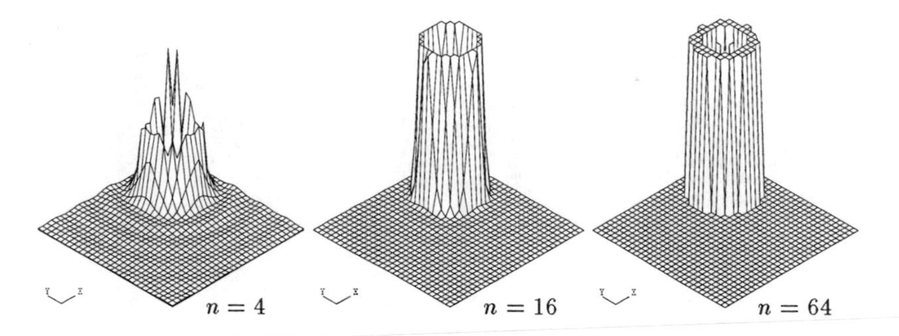

FIG. 7.2. *The reconstructed imaginary values of the contrast of the circular cylinder. The maximum reconstructed contrast is constrained to 1.2.*

contrast is larger than $\chi_{\max}$, the contrast is replaced by $\chi_{\max}$. In view of this modification, the residual errors have to be recomputed and the iterative scheme restarts with new contrast directions ξ_n. By enforcing the contrast gradients to be zero in all the points $\boldsymbol{p}$, where the contrast is equal to $\chi_{\max}$, the contrast directions vanish in these points and no updating of the contrast takes place in these points. Operating in this way, the scheme is able to 'concentrate' on updating the contrast at the remaining points. This accelerates the reconstruction and visualization of the boundary of the object. We illustrate this procedure for our example. We take $\chi_{\max} = i\,1.2$. We then solved the inverse problem with this upper limit and some surface plots of the reconstructed profiles (the imaginary part of the contrast χ) are presented in Fig. 7.2. The reconstructed boundary is very close to that of the circular cylinder.

7.2. Blind reconstruction from experimental data. The previous tests were performed with synthetic, i.e. computer simulated, scattering experiments and thus were not free from the possibility that they were tainted by an "inverse crime" of somehow using knowledge of the scatterer to favorably influence the reconstruction. To guarantee innocence of this "inverse crime", the inversion algorithm was tested using "real", experimental, data generated at the Ipswich test site of Rome Laboratory, Hanscom Air Force Base. The target was known to be symmetric about the x and y axes and lie inside a circle of radius 0.060 m, but was otherwise unspecified. Bistatic scattering measurements [8] were made in a plane perpendicular to the axis of a cylindrical object, 30 cm (10 λ) in length. The frequency of operation was 10 GHz (λ was 3 cm). The scattered fields were collected for eight incident angles, $\phi^{\text{inc}} = \{0°, 5°, 10°, 15°, 20°, 45°, 60°, 90°\}$ ($J = 8$), over an observation sector $0 \leq \phi^{\text{sct}} \leq 350°$ with a sample spacing $\Delta\phi^{\text{sct}} = 10°$. These so-called "Ipswich data" were used to reconstruct the unknown object. To obtain scattered-field data from incident waves distributed around the object, we took advantage of the a priori information

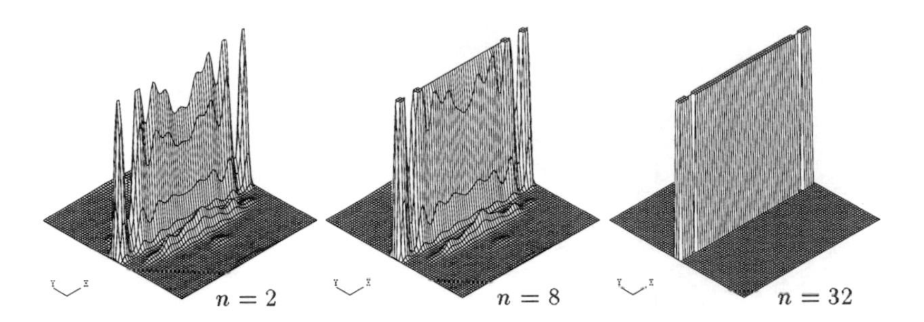

FIG. 7.3. *The reconstructed imaginary values of the contrast of the mystery object. The maximum reconstructed contrast is constrained to 1.0.*

that the mystery object is symmetric with respect to the planes $x = 0$ and $y = 0$. Doing so we obtain scattered-field data from 28 excitations ($J = 28$). We assume that the object is located inside a test square divided into 63×63 subsquares of 0.002×0.002 m^2. The results of the reconstruction are presented in Fig. 7.3. It shows that the mystery object is a strip of about a width of 12 cm and a thickness of less than or equal to 4 mm, which was later revealed to be quite accurate [24].

8. Comparison of TM case, TE case, and combined case. So-far, we have presented some numerical results using the modified gradient approach for the TM case in electromagnetics by using the object and data equations (1.6) – (1.7). The method has also been applied in the TE case in electromagnetics by using the object and data equations (1.13) – (1.14). Synthetic data were generated by numerical solution of the forward problem in both cases using a Galerkin method. Note that in the two cases different approximating subspaces have to be used: in the TM case, we use piecewise constant expansion functions for both the fields and the contrast χ; in the TE case we use piecewise linear expansion functions for the fields and piecewise constant expansion functions for the contrast $\mathcal{M}$. Both for comparison as well as to generate an algorithm combining

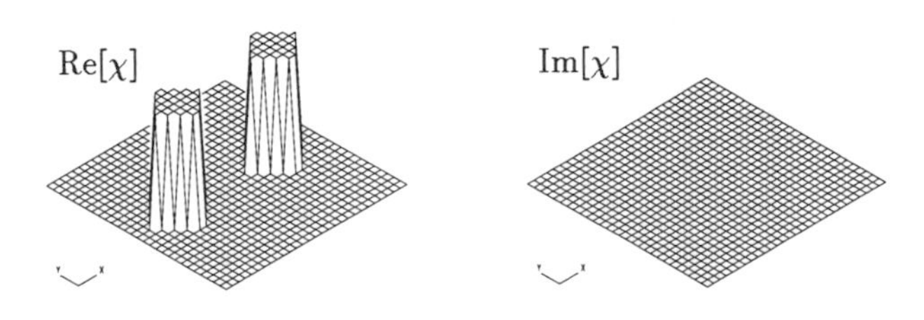

FIG. 8.1. *(a) The original profile.*

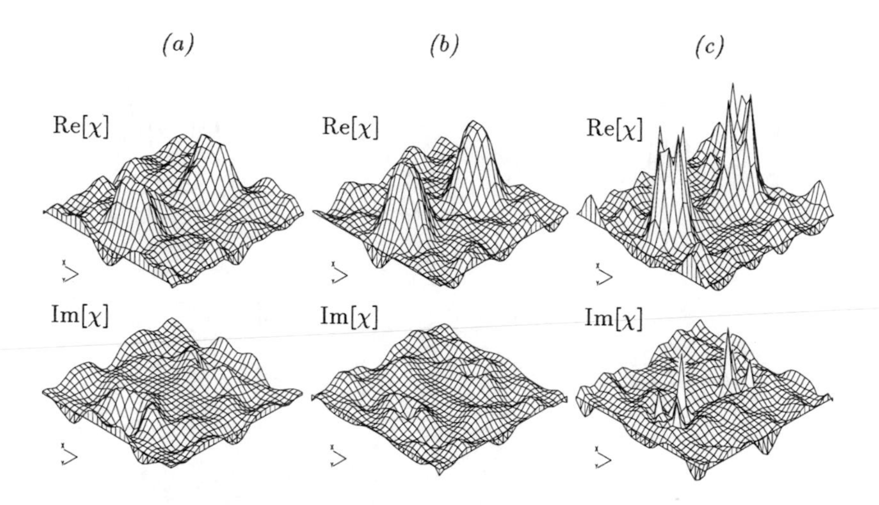

FIG. 8.2. *(a) The reconstructed profiles after 64 iterations, using 10 source/receiver stations; (a) TM case, (b) TE case, (c) combined case.*

both polarizations, we rewrite the equations for the TM case in such a way that the contrast χ is replaced by $\mathcal{M}$. That is, in the TM case we define renormalized field quantities, so that

$$\chi u_j = \mathcal{M} v_j \,, \tag{8.1}$$

in which case equations (1.9) and (1.10) become

$$u_j^{\text{sct}} - G_S \mathcal{M} v_j = 0 \,. \tag{8.2}$$

$$u_j^{\text{inc}} - v_j + \mathcal{M} v_j + G_D \mathcal{M} v_j = 0 \,, \tag{8.3}$$

These equations may be used to generate an alternate framework for a modified gradient solution of the TM case, while (1.15) and (1.16) may be used in a modified gradient solution in the TE case. Moreover, the two algorithms may be combined if scattered field data are available at both polarizations. In the latter case the cost functional to be minimized is the sum of the cost functionals of the TM and TE cases. The modified gradient method involves updating of the TE fields with one complex parameter and the TM fields with another complex parameter, while the contrast in both cases is updated with the same parameter. Hence, in each iteration of the modified gradient method we must solve a nonlinear problem in three complex parameters.

We illustrate the TE, TM and combined case in a numerical example in which the measurement surface S was chosen to be a circle of radius 9λ

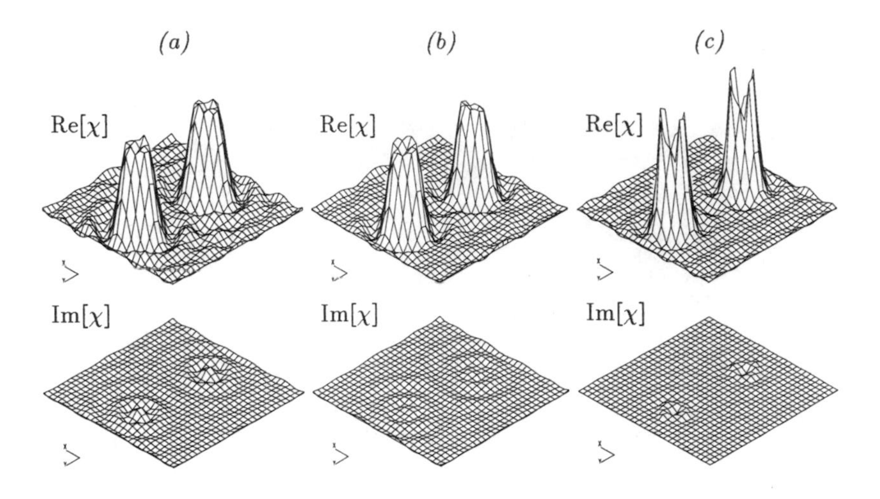

FIG. 8.3. *(a) The reconstructed profiles after 64 iterations, using 30 source/receiver stations; (a) TM case, (b) TE case, (c) combined case.*

containing the test domain D, which was a square, 3λ by 3λ, where λ is the wavelength in the background medium. A number of stations were located uniformly on the surface with each station serving successively as a line source and all stations acting as receivers. The test domain was discretized into 29×29 subsquares. The configuration to be reconstructed consists of two distinct homogeneous square cylinders of diameter $\frac{3}{4}\lambda$ with $\frac{3}{4}\lambda$ separation and contrast $\chi = 0.8$ ($\mathrm{Im}[\chi] = 0$), see Fig. 8.1. The reconstruction results ($\chi = \mathcal{M}/(1-\mathcal{M})$) are presented in Figs. 8.2 and 8.3 for 10 and 30 stations, respectively. For a small amount of scattering data, we observe that the TE case provides more resolution than the TM case, while the combined case improves the results as expected.

9. Total variation minimization. As a final example of the modified gradient method, we introduce the total variation as a penalty term. The idea of using such a constraint has been used in image enhancement, see [1,10,15,18,20], and is but one of many regularizing constraints that may be used for this purpose [3]. The form employed here combines the modified form suggested in [1] with a nonlinear constraint as in [9] and is described in more detail in [25]. The essential point is the replacement of the cost functional $F(\boldsymbol{u}, \chi)$ defined in (6.1) with a new cost functional

$$F'(\mathbf{u}, \chi, w_{TV}, \delta) := F(\mathbf{u}, \chi) + w_{TV} \int_D \sqrt{|\nabla\chi|^2 + \delta^2}\, dv\,. \tag{9.1}$$

The choice of the penalty parameter w_{TV}, which may be considered the reciprocal of the Lagrange multiplier for the problem of minimizing the total variation subject to the constraint $F = 0$, and the small parameter

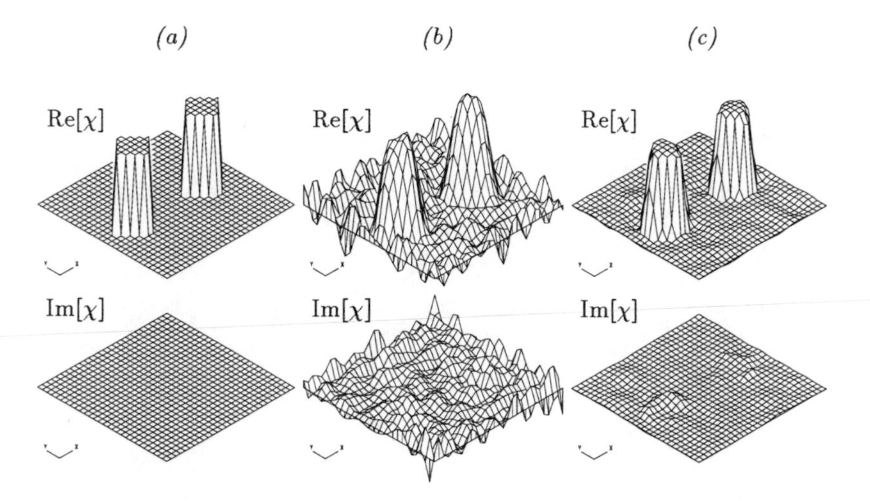

FIG. 9.1. *(a) The original profile and the reconstructed profiles after 64 iterations, (b) without constraint and (c) with constraint; 10% noise.*

δ, which restores differentiability to the total variation, are not determined by mathematical necessity nor physical reasoning. Rather, the choices are determined through numerical experimentation. In the numerical examples considered subsequently we used $\delta = 0.01$ and the algorithm seemed insensitive to changes of less than an order of magnitude. More critical is the choice of w_{TV} which was determined only through considerable numerical experimentation in the specific examples considered. With these parameters chosen, the algorithm is essentially the same as previously described with F' replacing F. The fields and the contrast are defined iteratively by (6.3) and the starting values are found exactly as previously described. The presence of the total variation (TV) term has no effect on the update directions for the field, but the gradient direction for the contrast has to be replaced by

$$g_n^{d'} = g_n^d - w_{TV} \nabla \cdot \left(\frac{\nabla \chi_{n-1}}{\sqrt{|\nabla \chi_{n-1}|^2 + \delta^2}} \right) . \tag{9.2}$$

The constants α_n and β_n are chosen to minimize the functional F' in which only α_n and β_n are unknown.

The addition of the total variation has a dramatic effect on the quality of the reconstructions. We illustrate this with three numerical examples which have been previously considered without the total variation term. The first example is the configuration used in the previous section (see Fig. 8.1. This is a "blocky" (piecewise constant) configuration of the class for which Dobson and Santosa [10] predict the total variation term should be effective. We choose to corrupt the data with additive white noise of

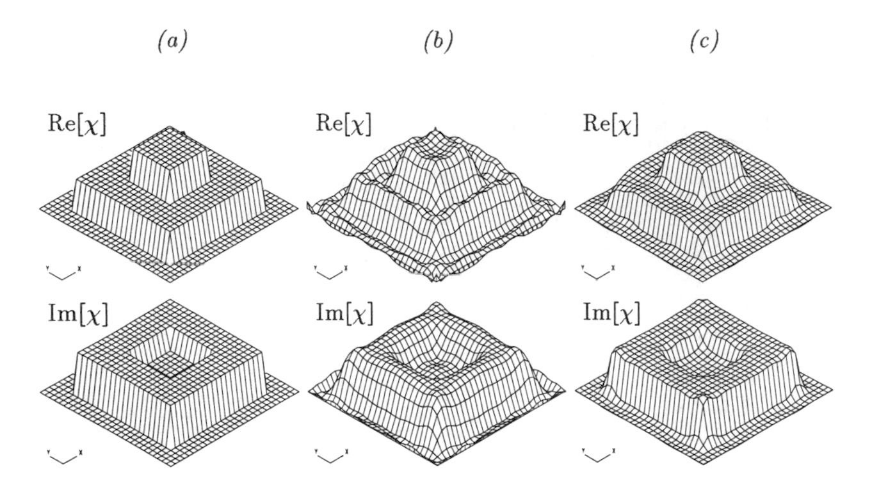

FIG. 9.2. *(a) The original profile and the reconstructed profiles after 128 iterations, (b) without constraint and (c) with constraint.*

10% of the maximum value of the measured data. The reconstruction after 64 iterations without and with the TV term are shown in Figs. 9.1b – 9.1c. While the TV enhanced reconstruction is manifestly superior, additional iterations makes the comparison even more striking. The unmodified reconstruction is relatively unchanged after more iterations while the TV enhanced reconstruction continues to improve. This is consistent with the original intent of the TV term to flatten oscillations due to noise.

The second example also involves a "blocky" contrast but this time with non-zero imaginary part. The object consisted of concentric square cylinders, an inner cylinder, λ by λ, with contrast $\chi = 0.6+2.0i$ surrounded by an outer cylinder, 2λ by 2λ, with contrast $\chi = 0.3 + 0.4i$ as shown in Fig. 9.2a. The reconstructions after 128 iterations are shown in Fig. 9.2b, without the TV term and Fig. 9.2c, with the TV term. As before the improvement in the reconstruction by the TV enhanced modified gradient method is apparent. Even more dramatic is the fact that while additional iterations have no visible effect on the unenhanced reconstruction, additional iterations of the enhanced reconstruction together with some tuning of the parameter w_{TV} produce a reconstruction that is extremely close to the actual profile.

As the third example we examined a smooth, non"blocky", contrast to see if the TV term, which tends to flatten oscillations would adversely affect the reconstruction which was rather good using the unenhanced method. We take the sinusoidal configuration described in the numerical example of Section 6 (see Figs. 6.1 and 9.3a). The reconstructions after 128 iterations are shown in Figs. 9.3b – 9.3c and it is clear that the TV term, rather than

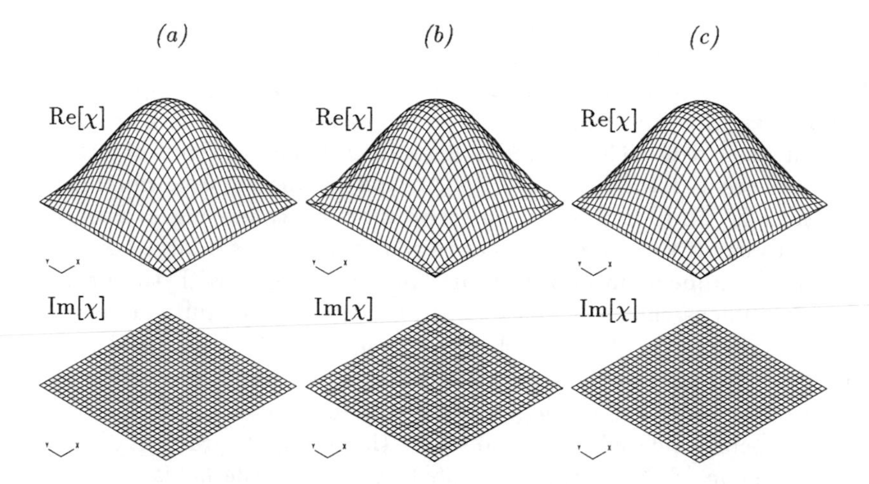

FIG. 9.3. *(a) The original profile and the reconstructed profiles after 128 iterations, (b) without constraint and (c) with constraint.*

affecting the reconstruction adversely, actually improves the quality both damping out unwanted oscillations in the imaginary part and more closely approximating the contrast at the boundary of the test domain.

10. Conclusions. In this paper we have reviewed a number of methods of solving the inverse problem of reconstructing the constitutive properties of a bounded two-dimensional inhomogeneity from values of the scattered field on a surface entirely exterior to the object, when the object is illuminated successively by known single-frequency incident fields originating in different points of the exterior domain. All of the methods discussed involve minimizing the discrepancy between the actual measurements and a representation of the scattered field on the surface of measurements, the data equation. An essential complication in the problem is due to the fact that not only are the constitutive parameters or contrast unknown but the fields inside the object are unknown as well. The difference in how the unknown interior fields are treated is what distinguishes the various approaches.

In the Born approximation the interior fields are simply approximated by the incident fields. In the iterative Born scheme the field is updated by solving the forward problem using the latest known update of the contrast. The contrast is updated by minimizing the error in the data equation by standard gradient based optimization methods. These methods work well when the contrast between the object and the background medium is small and deteriorate with increasing contrast. The distorted Born and Newton-Kantorovich methods, which were developed independently and only recently shown to be equivalent, are similar to the iterative Born method, but employ a more accurate approximation in the data equation. This extends the range of contrasts that can be reconstructed but has proven to be sensitive to noise in the data. Like the iterative Born approach, each iteration requires the solution of a forward or direct scattering problem for each incident field. This necessity is avoided if the data and object equations are treated together as a system and a functional combining both equations is minimized by an objective function consisting of both fields and contrast. If these unknowns are strung together as a large vector then standard nonlinear optimization methods may apply with the drawback that the dimensions of the discretized problem may be quite large.

The alternative described here in some detail is a modified gradient method in which the fields and the contrast are simultaneously updated either in the direction of the gradient of the functional with respect to the function being updated or, more often, in the conjugate gradient directions. In the version of the method presented here, all of the fields are updated in different directions with the same complex weight or step length and the contrast is also updated in a gradient or conjugate gradient direction with a different weight. These two weights are found solving the algebraic optimization problem of minimizing the cost functional which depends only on these two complex parameters. In this way the bilinear character of the cost functional is retained. The method was shown to be effective in both TM and TE electromagnetics. Both the real and imaginary parts of the contrast are simultaneously recovered and the approach was shown to be quite stable with respect to noise. The limits of reconstructibility were explored and it was shown how the inclusion of a priori information such as non-negativity of the imaginary part of the contrast could extend these limits. In fact the shape of impenetrable objects was demonstrated to be recoverable using this approach. The inclusion of the total variation as a penalty term was demonstrated to have a remarkable effect on improving the reconstructions.

Current and future work with this method includes the extension to three-dimensional problems, investigation of the effects of using frequency diverse incident waves, development of a systematic way of choosing the penalty parameter, methods of incorporating different kinds of a priori information into the cost functional, and improving the initial guess with which to begin the iterative process.

REFERENCES

[1] Acar, R., and Vogel, C.R., *Analysis of bounded variation penalty methods for ill-posed problems*, Inverse problems, **10** (1994), pp. 1217–1229

[2] Angell, T.S., Kleinman, R.E., and Roach, G.F., *An inverse transmission problem for the Helmholtz equation*, Inverse Problems, **3** (1986), pp. 149–180.

[3] Aubert, G., Barlaud, M., Blanc-Féraud, L., and Charbonnier, P., *Deterministic edge-preserving regularization in computed imaging*, IEEE Trans. Image Processing, to appear.

[4] BARBER, D., AND BROWN, B., *Applied potential tomography*, J. Phys. E: Sci. Instrument., **17** (1984), pp. 723–733.

[5] BARKESHLI, S., AND LAUTZENHEISER, R.G., *An iterative method for inverse scattering problems based on an exact gradient search*, Radio Science, **29** (1994), pp. 1119–1130.

[6] CHEW, W.C., AND WANG, Y.M., *Reconstruction of two-dimensional permittivity distribution using the distorted Born iterative method*, IEEE Trans. Med. Imag., **9** (1990), pp. 218–225.

[7] COLTON, D., AND KRESS, R., Inverse Acoustic and Electromagnetic Scattering Theory, Berlin Heidelberg: Springer-Verlag, 1992.

[8] COTÉ, M.G., *Automated swept-angle bistatic scattering measurements using continuous wave radar*, IEEE Trans. Instrum. Meas., **41** (1992), pp. 185–192.

[9] DOBSON, D.C., AND SANTOSA, F., *An image-enhancement technique for electrical impedance tomography*, Inverse Problems, **10** (1994), pp. 317–334.

[10] DOBSON, D.C., AND SANTOSA, F., *Recovery of blocky images from noisy and blurred data*, SIAM J. Appl. Math., **56** (1996), 1181–1198.

[11] FRANCHOIS, A., Contribution a La Tomographie Microonde: Algorithmes de Reconstruction Quantitative et Verifications Experimentales, These, Université de Paris-XI, 1993.

[12] KLEINMAN R.E., VAN DEN BERG, P.M., *A modified gradient method for two-dimensional problems in tomography*, Journal of Computational and Applied Mathematics, **42** (1992), pp. 17–35.

[13] KLEINMAN R.E., VAN DEN BERG, P.M., *An extended range modified gradient technique for profile inversion*, Radio Science, **28** (1993), pp. 877–884.

[14] KLEINMAN, R.E., AND VAN DEN BERG, P.M., *Two-dimensional location and shape reconstruction*, Radio Science, **29** (1994), pp. 1157–1169.

[15] OSHER, S., AND RUDIN, L., *Feature-oriented image enhancement using shock filters*, SIAM J. Numer. Anal., **27** (1990), pp. 919–940.

[16] OTTO, G.P., AND CHEW, W.C., *Inverse scattering of H_z waves using local shape-function imaging: a T-matrix formulation*, Int. J. Imag. Syst. Technol., **5** (1994), pp. 22–27.

[17] ROGER, A., *A Newton-Kantorovich algorithm applied to an electromagnetic inverse problem*, IEEE Trans. Antennas Propgat., **AP–29** (1981), pp. 232–238.

[18] RUDIN, L., OSHER, S., AND FATEMI, C., *Nonlinear total variation based noise removal algorithm*, Physica D, **60** (1992), pp. 259–268.

[19] SABBAGH, H.A., AND LAUTZENHEISER, R.G., *Inverse problems in electromagnetic nondestructive evaluation*, Int. Journal of Applied Electromagnetics in Materials, **3** (1993), pp. 253–261.

[20] SANTOSA, F., AND SYMES, W., *Reconstruction of blocky impedance profiles from normal-incidence reflection seismograms which are band-limited and miscalibrated*, Wave Motion, **10** (1988), pp. 209–230.

[21] SANTOSA, F., AND VOGELIUS, M., *A backpropagation algorithm for electrical impedance imaging*, SIAM J. Appl. Math., **50** (1990), pp. 216–243.

[22] TABBARA, W., DUCHÊNE, B., PICHOT, CH., LESSELIER. D., CHOMMELOUX, L., AND JOACHIMOWICZ, N., *Diffraction tomography: Contribution to the analysis of applications in microwaves and ultrasonics*, Inverse Problems, **4** (1988), pp. 305–331.

[23] TIJHUIS, A.G., *Born-type reconstruction of material parameters of an inhomogeneous lossy dielectric slab from reflected-field data*, Wave Motion, **11** (1989), pp. 151–173.

[24] VAN DEN BERG, P.M., COTÉ, M.G., AND KLEINMAN, R.E., *"Blind" shape reconstruction from experimental data*, IEEE Trans. Antennas and Propagat., **AP-43** (1995), pp. 1389–1396.

[25] VAN DEN BERG, P.M., AND KLEINMAN, R.E., *A total variation enhanced modified gradient algorithm for profile reconstruction*, Inverse Problems, **11** (1995), pp. L5–L10.

[26] WEN, L., KLEINMAN, R.E., AND VAN DEN BERG, P.M., *Modified gradient profile inversion using H-polarized waves*, Digest of the IEEE Antennas and Propagation Society International Symposium, Newport beach, California, June 18 – June 23 (1995), pp. 1598–1601.

[27] YORKEY, T., WEBSTER, J., AND TOMPKINS, W., *Comparing reconstruction algorithms for electrical impedance tomography*, IEEE Trans. Biomedical Eng., **BME–34** (1987), pp. 843–852.

ATMOSPHERIC DATA ASSIMILATION BASED ON THE REDUCED HESSIAN SUCCESSIVE QUADRATIC PROGRAMMING ALGORITHM

Y.F. XIE*

Abstract. Mesoscale weather forecasting cannot be improved until a better data assimilation is obtained. Four dimensional variational analysis (4DVAR) provides the most elegant framework for data assimilation. One of the most critical issues of applying 4DVAR to weather prediction is how efficiently these variational problems can be solved. We introduce the reduced Hessian SQP algorithm for these problems and obtain an adjoint reduced Hessian SQP method, which is quadratically convergent since the exact reduced Hessian is used.

Key words. atmospheric data assimilation, constrained optimization, four dimensional variational analysis, reduced Hessian successive quadratic programming algorithms

1. Introduction. Assimilating observed data into numerical weather forecasting models is becoming increasingly important in variety contexts of improving the weather forecasts. The four dimensional variational analysis (4DVAR) provides an elegant framework for atmospheric data analysis [1]. However, because of the demand of tremendous computing power to solve a 4VDAR problem for a realistic prediction model, developing efficient numerical algorithms becomes extremely important of applying this analysis to weather forecasting models.

Two classes of methods are currently used for solving 4DVAR problems: one, such as the L-BFGS method, uses the first-order derivatives [8] and the other, such as truncated Newton methods, uses the second-order ones [5]. Even though Zou et al. [10] showed that truncated Newton methods require far fewer iterations than the L-BFGS method, the total execution time of these methods is about the same because of using inexact Hessian vector products in obtaining truncated Newton steps. The adjoint truncated Newton (ATN) method, recently developed to compute the exact Hessian vector product, improves the efficiency of truncated Newton methods as shown in [9]. However, all of the current techniques for solving 4DVAR require integrations of the original nonlinear model because they are based upon an unconstrained optimization framework, i.e., they satisfy the constraints at all iterations. From constrained optimization point of view, the current methods for solving a 4DVAR are feasible methods. Integrations of nonlinear numerical models usually causes an instability problem, particularly for the cases with an un-initialized initial guess (for more discussion about initialization of a prediction model, see [1]).

To further improve the efficiency of truncated Newton methods and

* NOAA, ERL/FSL, The Forecasting Systems Laboratory, 325 Broadway, Boulder, CO 80303 and CIRA, Colorado State University.

avoid the nonlinear instability problems, we introduce an efficient method of solving 4DVAR problems based upon a constrained optimization framework. We expect that the new method have the following three important features. It

- is more efficient;
- solves the nonlinear stability problem.

The new method applies the reduced Hessian successive quadratic programming technique to 4DVAR problems, which means we treat the 4DVAR as a constrained optimization problem instead of an unconstrained one. Using this new framework to implement truncated Newton methods, an exact Hessian vector product can be computed by integrating the linearized and adjoint models. Since it is not necessary to satisfy the nonlinear constraints at each iteration, it may intuitively converge even faster than feasible methods. As the reduced Hessian SQP method solves linearized constraints at each iteration, nonlinear instability problems are avoided.

In the following sections, we discuss how to apply the reduced Hessian technique to data assimilation problems. We describe the 4DVAR problem in Section 2 and derive the adjoint model in section 3. In Section 4, we introduce the reduced Hessian SQP method to 4DVAR problems and discuss the computation complexity. Finally we give comments on Section 5.

2. Four dimensional variational analysis demonstrated by the Burger equation. We discuss general discretized systems representative of general weather prediction models. However, for simplicity, we consider fluid which is governed by an equation $u_t + uu_x = 0$ where u is the velocity of the fluid and x and t are space and time variables, respectively. Note that even though we discuss a simple system, the analysis of the new technique works for general weather prediction models.

Let us consider how 4DVAR assimilates observed data into a numerical weather prediction model. Suppose over a domain $[a,b] \times [t',t'']$, there is a set of observed data $\{u^{obs}(x_i,t_i)\}_{i=1}^{I}$ with a set of weights $\{W_i\}_{i=1}^{I}$. The 4DVAR problem is to find a function u^* satisfying the dynamical system such that

$$\begin{aligned} f(u^*) \quad &= \quad \text{minimize} \quad \sum_{i=1}^{I} W_i[u(x_i,t_i) - u^{obs}(x_i,t_i)]^2 \\ &\quad\quad \text{subject to} \quad u_t + uu_x = 0. \end{aligned}$$

For simplicity, the initial and boundary condition $u(a,t)$, $u(b,t)$ and $u(x,0)$ are considered as control variables. To further simplify our discussion, we will assume periodic boundary conditions for the remaining discussion. If a leap-frog scheme is used in time and a second-order centered finite difference is used in space, the variational problem can be rewritten as

$$\text{(2.1)} \quad \text{minimize} \qquad f(u) = \sum_{i=1}^{I} W_i[u(x_i,t_i) - u^{obs}(x_i,t_i)]^2$$

$$(2.2) \quad \text{subject to} \quad c(u) = u_j^{n+1} - u_j^{n-1} + u_j^n \frac{\Delta t}{\Delta x}\left(u_{j+1}^n - u_{j-1}^n\right) = 0,$$
$$j = 2, 3, \cdots, K, \quad \text{and} \quad n = 2, 3, \cdots, M+1.$$

where $K+1$ is the number of grid points used in the x direction and $M+1$ is the number of time steps taken over $[t', t'']$. Variables $\{u_1^n\}_{n=3}^{M+1}$ and $\{u_{K+1}^n\}_{n=3}^{M+1}$ are determined by the boundary conditions. This is a typical constrained optimization problem if we consider the state variables at all grid points as unknowns.

There are many methods to approximately solve 4DVAR problems. The different methods for Four Dimensional Data Assimilation (FDDA) problems are related to the ones for solving 4DVAR problems by Lorenc [4]; some examples include function fitting, successive correction, nudging, statistical interpolations and Kalman-Bucy filter. All of these methods approximately solve 4DVAR problems in different degrees. As Daley [1] pointed out, the present methods cannot be "*capable of assimilating the future data into the future models.*"

Recently as computing capability has greatly expanded, many advanced techniques are being applied to 4DVAR problems. The current computational framework is to use the constraint (2.2) to reduce the 4DVAR problem to an unconstrained optimization problem such that the descent methods can be applied through the adjoint models. For example, assume u is periodic in x for simplicity and then for any initial conditions, $\{u_j^1\}_{k=1}^{K+1}$ and $\{u_j^2\}_{k=1}^{K+1}$, the objective function f is determined after the integration of the constraint (2.2). That is, the initial conditions are the control variables and f can be viewed as a function of initial conditions. Solving (2.1)-(2.2) is equivalent to minimizing the function f with respect to the control variables. This, therefore, becomes an unconstrained optimization problem and then the unconstrained optimization techniques can be applied as long as the gradients of the objective function in terms of the control variables are available, e.g., the steepest descent direction methods [2], the conjugate-gradient methods [6] and the Newton-like methods [10]. The adjoint model of optimal control theory is introduced to evaluate the gradients efficiently. Lewis and Derber [2] described the adjoint model in detail. Thus an adjoint model and the unconstrained optimization techniques basically form the current computational framework.

3. Formulation of the adjoint equations. To show how the reduced Hessian SQP method can be implemented efficiently, we derive the definition of the generalized adjoint model for a given discretized model.

Since we evaluate the first-order derivatives, a linearized version of the original model will be sufficient. For any given state, say $\bar{u}$, the linearized constraint (2.2) is

$$\begin{aligned} \bar{u}_j^{n+1} + d_j^{n+1} - \bar{u}_j^{n-1} - d_j^{n-1} + \frac{\Delta t}{\Delta x}[\bar{u}_j^n(\bar{u}_{j+1}^n - \bar{u}_{j-1}^n) \\ +\bar{u}_j^n(d_{j+1}^n - d_{j-1}^n) + d_j^n(\bar{u}_{j+1}^n - \bar{u}_{j-1}^n)] \quad &= \quad 0 \end{aligned}$$

where $d = u - \bar{u}$. Ordering the unknowns u_j^n as a vector

$$u = (u_K^{M+1}, \cdots, u_1^{M+1}, \cdots, u_K^n, \cdots, u_1^n, \cdots, u_K^2, \cdots, u_1^2, u_K^1, \cdots, u_1^1),$$

the linearized system can be written as

$$c(\bar{u}) + A^T(\bar{u})(u - \bar{u}) = A_B^T u_B + A_N^T u_N - b = 0, \tag{3.1}$$

where b is a constant vector for given $\bar{u}$, T stands for transpose and

$$u_B = (u_K^{M+1}, \cdots, u_1^{M+1}, \cdots, u_K^n, \cdots, u_1^n, \cdots, u_K^3, \cdots, u_1^3)$$
$$u_N = (u_K^2, \cdots, u_1^2, u_K^1, \cdots, u_1^1)$$

and $A^T = (A_B^T \;\; A_N^T)$ is a decomposition corresponding to u_B and u_N. It is not difficult to verify that A_B^T could be written as,

$$A_B^T = \begin{pmatrix} I & D^M & -I & 0 & 0 & \cdots & 0 \\ 0 & I & D^{M-1} & -I & 0 & \cdots & 0 \\ \cdots & & \cdots & \cdots & \cdots & & \cdots \\ 0 & \cdots & 0 & I & D^5 & -I & 0 \\ 0 & \cdots & 0 & 0 & I & D^4 & -I \\ 0 & \cdots & 0 & 0 & 0 & I & D^3 \\ 0 & \cdots & 0 & 0 & 0 & 0 & I \end{pmatrix} \tag{3.2}$$

where I is a $K \times K$ identity matrix,

$$D^n = \begin{pmatrix} c_K^n & -b_K^n & 0 & \cdots & 0 \\ b_{K-1}^n & c_{K-1}^n & -b_{K-1}^n & \cdots & 0 \\ \cdots & & \cdots & & \cdots \\ 0 & \cdots & b_2^n & c_2^n & -b_2^n \\ 0 & \cdots & 0 & b_1^n & c_1^n \end{pmatrix}$$

where $c_i^n = \frac{\Delta t}{\Delta x}(\bar{u}_{i+1}^n - \bar{u}_{i-1}^n)$ and $b_i^n = \frac{\Delta t}{\Delta x}\bar{u}_i^n$. A_B^T is an upper triangular nonsingular matrix. The gradient of f with respect to u_N can be computed by the chain rule and it has the form of

$$\frac{df}{du_N} = \left(\frac{\partial u_B}{\partial u_N}\right)^T \nabla_{u_B} f + \nabla_{u_N} f.$$

where $\nabla_{u_B} f = \frac{\partial f}{\partial u_B}$ and $\nabla_{u_N} f = \frac{\partial f}{\partial u_N}$. Equation (3.1) implies $\frac{\partial u_B}{\partial u_N} = -A_B^{-T} A_N^T$ and

$$\frac{df}{du_N} = \nabla_{u_N} f - A_N A_B^{-1} \nabla_{u_B} f. \tag{3.3}$$

Assume

$$y = A_B^{-1} \nabla_{u_B} f$$

and then

$$A_B y = \nabla_{u_B} f. \tag{3.4}$$

Because A_B is a lower triangular matrix, one forward-substitution gives the vector y, and the computation complexity is the same as the integration of the original system. Or more clearly, (3.4) implies that for any interior points (i, n), we have

$$\begin{aligned} \frac{y_i^{n-1} - y_i^{n+1}}{\Delta t} &+ \frac{\bar{u}_{i+1}^{n-1} - \bar{u}_{i-1}^{n-1}}{\Delta x} y_i^n + \\ &+ \frac{\bar{u}_{i-1}^{n-1}}{\Delta x} y_{i-1}^n - \frac{\bar{u}_{i+1}^{n-1}}{\Delta x} y_{i+1}^n = \frac{(\nabla_{u_B} f)_{i,n}}{\Delta t} \end{aligned}$$

where $(\nabla_{u_B} f)_{i,n}$ stands for the corresponding component of the grid point (i, n). If let $\Delta t \to 0$ and $\Delta x \to 0$, this system approximates to $-y_t - \bar{u} y_x$ because

$$\begin{aligned} &\frac{y_i^{n-1} - y_i^{n+1}}{\Delta t} \longrightarrow -y_t; \qquad \frac{\bar{u}_{i+1}^{n-1} - \bar{u}_{i-1}^{n-1}}{\Delta x} y_i^n \longrightarrow \bar{u}_x y; \\ &\frac{\bar{u}_{i-1}^{n-1}}{\Delta x} y_{i-1}^n - \frac{\bar{u}_{i+1}^{n-1}}{\Delta x} y_{i+1}^n \\ &= \frac{\bar{u}_{i-1}^{n-1}}{\Delta x}(y_{i-1}^n - y_{i+1}^n) + \frac{\bar{u}_{i-1}^{n-1} - \bar{u}_{i+1}^{n-1}}{\Delta x} y_{i+1}^n \longrightarrow -\bar{u} y_x - \bar{u}_x y. \end{aligned}$$

To solve (3.4) is essentially to integrate the original model backward in time. We call (3.4) the adjoint equation of the discretized problem (3.1) because A_B is the adjoint operator of A_B^T. Notice that for any right-hand side, y can be solved similarly. After computing y, the evaluation of the gradient $\frac{df}{du_N}$ becomes trivial.

Interestingly, this phenomenon occurs for not only this particular equation and integration scheme but also any weather prediction model with different discretization schemes, for examples, finite difference, finite element, semi-Lagrangian and spectral models. Instead of the scalar upper triangular form, A_B^T is in general a blocked upper triangular matrix compared with (3.2). That is,

$$A_B^T = \begin{pmatrix} A_{11} & * & * & \cdots & * \\ 0 & A_{22} & * & \cdots & * \\ \cdots & & \cdots & & \cdots \\ 0 & 0 & \cdots & 0 & A_{MM} \end{pmatrix}$$

where A_{ii} for $i = 1, \cdots, M$ are square matrices and may or may not be identity matrices whose dimensions are less than the number of the function values over the grid points at any given time. For example, A_{ii} is a banded matrix for the finite element methods [12]. It is not difficult to verify that the finite difference, finite element, semi-Lagrangian and spectral models can be written as this formulation ([3], [7] and [12]).

Based on this definition of the adjoint model, we offer the following comments.

- No matter what dynamical system is used as the strong constraint, the system (3.4) can be solved as efficiently as the integration of original system, which is essentially integrating the original system backward in time.
- Constrained optimization algorithms can be implemented to solve 4DVAR problems by using different right hand side for (3.4) instead of using the constraints to reduce them to unconstrained problems.

Now we consider implementation of a new computational framework by using this new definition of an adjoint model.

4. A new computational framework for 4DVAR. Even though an adjoint model evaluates gradients efficiently, the methods using gradient information only is not satisfactory. Instead of using the constraint to reduce (2.1)-(2.2) to an unconstrained optimization problem, a new computational framework treats the 4DVAR problem as a constrained optimization problem,

$$\min \quad f(u) \tag{4.1}$$

$$\text{s.t.} \quad c(u) = 0. \tag{4.2}$$

A new computational framework applies the reduced Hessian successive quadratic programming (RHSQP, see [11]) method to (4.1)-(4.2). Now we briefly review the reduced Hessian SQP method.

For the constrained problem, the Lagrangian function is given by

$$L(u, \lambda) = f(u) + \lambda c(u)$$

where λ is a Lagrangian multiplier. Suppose at iteration k there is an approximation to the solution of (4.1)-(4.2), say u_k. Let B_k be the exact reduced Hessian matrix, i.e.,

$$B_k = Z_k^T \nabla_{uu} L(u_k, \lambda_k) Z_k$$

where

$$\lambda_k = -(A_k)_L^{-1} \nabla_u f,$$

$(A_k)_L^{-1}$ is a left inverse matrix of A_k and Z_k is a basis matrix of the null space of A_k^T (i.e., $A_k^T Z_k = 0$). Using the current approximation u_k, a quadratic programming is constructed

$$\min \quad \nabla f_k^T d + \frac{1}{2} d^T (Z_k)_L^{-T} B_k (Z_k)_L^{-1} d \tag{4.3}$$

$$\text{s.t.} \quad c(u_k) + A_k^T d = 0 \tag{4.4}$$

where $(Z_k)_L^{-1}$ is a left inverse matrix of Z_k. Its solution may be expressed as

$$d_k = h_k + v_k$$

where $h_k = Z_k s_k$ and s_k satisfies

$$B_k s_k = -Z_k^T \nabla f_k \tag{4.5}$$

and

$$v_k = -(A_k)_L^{-T} c_k. \tag{4.6}$$

By computing d_k, a new approximation of the solution can be obtained

$$u_{k+1} = u_k + d_k.$$

This gives the general reduced Hessian SQP method and this method is quadratically convergent because the exact Hessian is used.

For a 4DVAR problem, the computation of h_k (or s_k) and v_k can be simplified as integrations of the linearized model and its adjoint model. Suppose

$$A_k = \begin{pmatrix} A_B \\ A_N \end{pmatrix}$$

and we choose a basis matrix Z_k and the inverse matrices as:

$$Z_k = \begin{pmatrix} -A_B^{-T} A_N^T \\ I \end{pmatrix} \tag{4.7}$$

$$(A_k)_L^{-1} = (A_B^{-1} \quad 0) \tag{4.8}$$

$$(Z_k)_L^{-1} = (0 \quad I)\,. \tag{4.9}$$

Thus, the vertical step v_k and the Lagrangian multiplier can be computed by integrating the linearized model and its adjoint model, respectively. For example, the vertical step v_k can be computed by the following decomposition

$$v_k = \begin{pmatrix} (v_k)_B \\ (v_k)_N \end{pmatrix} = \begin{pmatrix} A_B^{-T} c_k \\ 0 \end{pmatrix}.$$

The remaining question is how to compute h_k or s_k.

Now let us consider the computation of s_k, for instance. For a 4DVAR problem, the matrix B_k is so large that it is impossible to use a direct method to invert B_k to obtain s_k by (4.5). Iterative methods are usually applied to (4.5) in obtaining a truncated Newton step and these iterations are called the inner iterations. Each inner iteration usually requires a Hessian and vector product only by using, for example, the conjugate gradient method. Let us look at the expense computing an exact Hessian vector product under this new computational framework. For any given vector w, computation of a Hessian vector product, $B_k w$, can be decomposed into the following steps

$$w^1 = Z_k w, \quad w^2 = \nabla_{uu} L(u_k, \lambda_k) w^1, \quad \text{and} \quad w^3 = Z^T w^2$$

and w^3 is the product, $B_k w$. First,

$$w^1 = \begin{pmatrix} w_B^1 \\ w_N^1 \end{pmatrix} = Z_k w = \begin{pmatrix} -A_B^{-T} A_N^T w \\ w \end{pmatrix}.$$

One forward integration of the linearized model, the system $A_B^T w_B^1 = A_N^T(-w)$ can be solved. To compute w^2, consider the full Hessian

$$\nabla_{uu} L(u_k, \lambda_k) = \nabla^2 f_k + \sum_{i=1}^{|B|} (\lambda_k)_i \nabla^2 (c_k)_i$$

where $(\cdot)_i$ indicates a vector's i-th element and $|B|$ is the cardinality of the index set B corresponding to the basic variables, i.e., the state variables at all of interior points for this particular problem. For a 4DVAR problem, the number of nonzeros of this matrix is no more than the ones appearing in the linearized models and they can be computed explicitly using the objective function f and the gradient of the constraints A_B and A_N. This implies that the computation of w^2 is no more expensive than an integration of the linearized models. Finally,

$$w^3 = Z^T w^2 = (w^2)_N - A_N A_B^{-1} (w^2)_B$$

which costs one integration of the adjoint model in obtaining $A_B^{-1}(w^2)_B$. Therefore, for each exact Hessian vector product, it takes a period of time which is equivalent to three integrations of the linearized and adjoint models.

From the above discussion, we can see that this method requires only the linearized constraints and it is not necessary to integrate the nonlinear models. This is an important feature for 4DVAR problems because the instability problems may effect the applicability of the feasible methods to a 4DVAR problem and the instability problems are usually inevitable for uninitialized initial fields of a prediction model. Since reducing a constrained problem to an unconstrained one causes a higher nonlinearity of the objective function, applying the reduced Hessian SQP to 4DVAR problems could be more efficient comparing to the methods based on an unconstrained optimization framework. These two advantages,

- No nonlinear instability problem and
- Quadratic convergence rate,

make the adjoint RHSQP method one of the best candidates for atmospheric data assimilation.

5. Comments. The analysis of the adjoint reduced Hessian successive quadratic programming method shows its great potential in the atmospheric data assimilation. However there will be several important issues to be studied in the implementation and numerical experiments of the new technique, such as selection of a good starting point and development a good preconditioning scheme for the reduced Hessian matrix. There are some other techniques can be used to provide a good starting point, for instance, statistical interpolation. Forecasts in the past may be a good starting point too. We will discuss these issues in our numerical experiments.

Numerical experiments of this new method is being carried at the Forecasting Systems Laboratory (FSL), National Oceanic and Atmospheric Administration (NOAA) at Boulder. A non-hydrostatic model using the FSL CONUS C grid in space and the third order Adams-Bashforth in time is to be used as the dynamical constraints and a set of wind profiler data and ACAR data is to be fitted. We will report the numerical results when they are completed.

Acknowledgements The author thanks Dr. O. Talagrand and Prof. J. Nocedal for their invaluable communication on atmospheric data assimilation. The author also thanks the anonymous reviewer for his helpful comments and suggestions.

REFERENCES

[1] Roger Daley. *Atmospheric Data Analysis.* Cambridge University Press, 1991.

[2] J. Lewis and J. Derber. The use of adjoint equations to solve variational adjustment problems with advective constraints. *Tellus*, 37:309–327, 1985.

[3] Yong Li, I. M. Navon, Weiyu Yang, Xiaolei Zou, J. R. Bates, S. Moorthi, and R. W. Higgins. Variational data assimilation experiments with a multilevel semi-Lagrangian semi-implicit general circulation model. *Mon. Wea. Rev.*, 122:966–983, 1994.

[4] A. C. Lorenc. Analysis methods for numerical weather prediction. *Quart. J. R. Met. Soc.*, 112:1177–1194, 1986.

[5] S. G. Nash. Solving nonlinear programming problems using truncated Newton techniques. Numerical Optimization, pages 119–136. SIAM, Philadelphia, 1984.

[6] I. M. Navon and D. Legler. Conjugate-gradient methods for large-scale minimization in meteorology. *Mon. Wea. Rev.*, 115:1749–1502, 1987.

[7] I. M. Navon, X. Zou, J. Derber, and J. Sela. Variational data assimilation with an adiabatic version of the NMC spectral model. *Mon. Wea. Rev.*, 120:1433–1446, 1992.

[8] J. Nocedal. Updating quasi-Newton matrices with limited storage. *Math. Comput.*, 35:773–782, 1980.

[9] Z. Wang, I. M. Navon, X. Zou, and F. X. LeDimet. The adjoint truncated Newton algorithm for large-scale unconstrained optimization. *Computational Optimization and Applications*, 4 (3), 241–262.

[10] Zou X, I. M. Navon, M. Berger, K. H. Phua, T. Schlick, and F. X. Le Dimet. Numerical experience with limited-memory quasi-Newton and truncated Newton methods. *SIAM J. Optimization*, 3(3):582–608, 1993.

[11] Y. F. Xie. The reduced Hessian successive quadratic programming algorithms for solving large scale equality constrained optimization probelm. *Ph.D Dissertation,* University of Colorado at Boulder, 1991.

[12] Keyun Zhu, I. M. Navon, and X. Zou. Variational data assimilation with variable resolution finite-element shallow water equations model. *Mon. Wea. Rev.*, 122(5):946–965, 1994.

IMA SUMMER PROGRAMS

1987 Robotics
1988 Signal Processing
1989 Robustness, Diagnostics, Computing and Graphics in Statistics
1990 Radar and Sonar (June 18 - June 29)
New Directions in Time Series Analysis (July 2 - July 27)
1991 Semiconductors
1992 Environmental Studies: Mathematical, Computational, and Statistical Analysis
1993 Modeling, Mesh Generation, and Adaptive Numerical Methods for Partial Differential Equations
1994 Molecular Biology
1995 Large Scale Optimizations with Applications to Inverse Problems, Optimal Control and Design, and Molecular and Structural Optimization
1996 Emerging Applications of Number Theory
1997 Statistics in Health Sciences
1998 Coding and Cryptography

SPRINGER LECTURE NOTES FROM THE IMA:

The Mathematics and Physics of Disordered Media
Editors: Barry Hughes and Barry Ninham
(Lecture Notes in Math., Volume 1035, 1983)

Orienting Polymers
Editor: J.L. Ericksen
(Lecture Notes in Math., Volume 1063, 1984)

New Perspectives in Thermodynamics
Editor: James Serrin
(Springer-Verlag, 1986)

Models of Economic Dynamics
Editor: Hugo Sonnenschein
(Lecture Notes in Econ., Volume 264, 1986)

The IMA Volumes in Mathematics and its Applications

Current Volumes:

23 **Signal Processing Part II: Control Theory and Applications of Signal Processing** L. Auslander, F.A. Grünbaum, J.W. Helton, T. Kailath, P. Khargonekar, and S. Mitter (eds.)
24 **Mathematics in Industrial Problems, Part 2** A. Friedman
25 **Solitons in Physics, Mathematics, and Nonlinear Optics** P.J. Olver and D.H. Sattinger (eds.)
26 **Two Phase Flows and Waves** D.D. Joseph and D.G. Schaeffer (eds.)
27 **Nonlinear Evolution Equations that Change Type** B.L. Keyfitz and M. Shearer (eds.)
28 **Computer Aided Proofs in Analysis** K. Meyer and D. Schmidt (eds.)
29 **Multidimensional Hyperbolic Problems and Computations** A. Majda and J. Glimm (eds.)
30 **Microlocal Analysis and Nonlinear Waves** M. Beals, R. Melrose, and J. Rauch (eds.)
31 **Mathematics in Industrial Problems, Part 3** A. Friedman
32 **Radar and Sonar, Part I** R. Blahut, W. Miller, Jr., and C. Wilcox
33 **Directions in Robust Statistics and Diagnostics: Part I** W.A. Stahel and S. Weisberg (eds.)
34 **Directions in Robust Statistics and Diagnostics: Part II** W.A. Stahel and S. Weisberg (eds.)
35 **Dynamical Issues in Combustion Theory** P. Fife, A. Liñán, and F.A. Williams (eds.)
36 **Computing and Graphics in Statistics** A. Buja and P. Tukey (eds.)
37 **Patterns and Dynamics in Reactive Media** H. Swinney, G. Aris, and D. Aronson (eds.)
38 **Mathematics in Industrial Problems, Part 4** A. Friedman
39 **Radar and Sonar, Part II** F.A. Grünbaum, M. Bernfeld, and R.E. Blahut (eds.)
40 **Nonlinear Phenomena in Atmospheric and Oceanic Sciences** G.F. Carnevale and R.T. Pierrehumbert (eds.)
41 **Chaotic Processes in the Geological Sciences** D.A. Yuen (ed.)
42 **Partial Differential Equations with Minimal Smoothness and Applications** B. Dahlberg, E. Fabes, R. Fefferman, D. Jerison, C. Kenig, and J. Pipher (eds.)
43 **On the Evolution of Phase Boundaries** M.E. Gurtin and G.B. McFadden
44 **Twist Mappings and Their Applications** R. McGehee and K.R. Meyer (eds.)
45 **New Directions in Time Series Analysis, Part I** D. Brillinger, P. Caines, J. Geweke, E. Parzen, M. Rosenblatt, and M.S. Taqqu (eds.)

46 **New Directions in Time Series Analysis, Part II**
D. Brillinger, P. Caines, J. Geweke, E. Parzen, M. Rosenblatt, and M.S. Taqqu (eds.)

47 **Degenerate Diffusions**
W.-M. Ni, L.A. Peletier, and J.-L. Vazquez (eds.)

48 **Linear Algebra, Markov Chains, and Queueing Models**
C.D. Meyer and R.J. Plemmons (eds.)

49 **Mathematics in Industrial Problems, Part 5** A. Friedman

50 **Combinatorial and Graph-Theoretic Problems in Linear Algebra**
R.A. Brualdi, S. Friedland, and V. Klee (eds.)

51 **Statistical Thermodynamics and Differential Geometry of Microstructured Materials**
H.T. Davis and J.C.C. Nitsche (eds.)

52 **Shock Induced Transitions and Phase Structures in General Media** J.E. Dunn, R. Fosdick, and M. Slemrod (eds.)

53 **Variational and Free Boundary Problems**
A. Friedman and J. Spruck (eds.)

54 **Microstructure and Phase Transitions**
D. Kinderlehrer, R. James, M. Luskin, and J.L. Ericksen (eds.)

55 **Turbulence in Fluid Flows: A Dynamical Systems Approach**
G.R. Sell, C. Foias, and R. Temam (eds.)

56 **Graph Theory and Sparse Matrix Computation**
A. George, J.R. Gilbert, and J.W.H. Liu (eds.)

57 **Mathematics in Industrial Problems, Part 6** A. Friedman

58 **Semiconductors, Part I**
W.M. Coughran, Jr., J. Cole, P. Lloyd, and J. White (eds.)

59 **Semiconductors, Part II**
W.M. Coughran, Jr., J. Cole, P. Lloyd, and J. White (eds.)

60 **Recent Advances in Iterative Methods**
G. Golub, A. Greenbaum, and M. Luskin (eds.)

61 **Free Boundaries in Viscous Flows**
R.A. Brown and S.H. Davis (eds.)

62 **Linear Algebra for Control Theory**
P. Van Dooren and B. Wyman (eds.)

63 **Hamiltonian Dynamical Systems: History, Theory, and Applications**
H.S. Dumas, K.R. Meyer, and D.S. Schmidt (eds.)

64 **Systems and Control Theory for Power Systems**
J.H. Chow, P.V. Kokotovic, R.J. Thomas (eds.)

65 **Mathematical Finance**
M.H.A. Davis, D. Duffie, W.H. Fleming, and S.E. Shreve (eds.)

66 **Robust Control Theory** B.A. Francis and P.P. Khargonekar (eds.)

67 **Mathematics in Industrial Problems, Part 7** A. Friedman

68 **Flow Control** M.D. Gunzburger (ed.)

69	**Linear Algebra for Signal Processing** A. Bojanczyk and G. Cybenko (eds.)
70	**Control and Optimal Design of Distributed Parameter Systems** J.E. Lagnese, D.L. Russell, and L.W. White (eds.)
71	**Stochastic Networks** F.P. Kelly and R.J. Williams (eds.)
72	**Discrete Probability and Algorithms** D. Aldous, P. Diaconis, J. Spencer, and J.M. Steele (eds.)
73	**Discrete Event Systems, Manufacturing Systems, and Communication Networks** P.R. Kumar and P.P. Varaiya (eds.)
74	**Adaptive Control, Filtering, and Signal Processing** K.J. Åström, G.C. Goodwin, and P.R. Kumar (eds.)
75	**Modeling, Mesh Generation, and Adaptive Numerical Methods for Partial Differential Equations** I. Babuska, J.E. Flaherty, W.D. Henshaw, J.E. Hopcroft, J.E. Oliger, and T. Tezduyar (eds.)
76	**Random Discrete Structures** D. Aldous and R. Pemantle (eds.)
77	**Nonlinear Stochastic PDEs: Hydrodynamic Limit and Burgers' Turbulence** T. Funaki and W.A. Woyczynski (eds.)
78	**Nonsmooth Analysis and Geometric Methods in Deterministic Optimal Control** B.S. Mordukhovich and H.J. Sussmann (eds.)
79	**Environmental Studies: Mathematical, Computational, and Statistical Analysis** M.F. Wheeler (ed.)
80	**Image Models (and their Speech Model Cousins)** S.E. Levinson and L. Shepp (eds.)
81	**Genetic Mapping and DNA Sequencing** T. Speed and M.S. Waterman (eds.)
82	**Mathematical Approaches to Biomolecular Structure and Dynamics** J.P. Mesirov, K. Schulten, and D. Sumners (eds.)
83	**Mathematics in Industrial Problems, Part 8** A. Friedman
84	**Classical and Modern Branching Processes** K.B. Athreya and P. Jagers (eds.)
85	**Stochastic Models in Geosystems** S.A. Molchanov and W.A. Woyczynski (eds.)
86	**Computational Wave Propagation** B. Engquist and G.A. Kriegsmann (eds.)
87	**Progress in Population Genetics and Human Evolution** P. Donnelly and S. Tavaré (eds.)
88	**Mathematics in Industrial Problems, Part 9** A. Friedman
89	**Multiparticle Quantum Scattering With Applications to Nuclear, Atomic and Molecular Physics** D.G. Truhlar and B. Simon (eds.)
90	**Inverse Problems in Wave Propagation** G. Chavent, G. Papanicolau, P. Sacks, and W.W. Symes (eds.)
91	**Singularities and Oscillations** J. Rauch and M. Taylor (eds.)

92 **Large-Scale Optimization with Applications, Part I: Optimization in Inverse Problems and Design**
L.T. Biegler, T.F. Coleman, A.R. Conn, F.N. Santosa (eds).

93 **Large-Scale Optimization with Applications, Part II: Optimal Design and Control**
L.T. Biegler, T.F. Coleman, A.R. Conn, F.N. Santosa (eds).

94 **Large-Scale Optimization with Applications, Part III: Molecular Structure and Optimization**
L.T. Biegler, T.F. Coleman, A.R. Conn, F.N. Santosa (eds).